# Cordyceps: Treating Diabetes, Cancer and Other Illnesses

## It could save your life

Nathalie Valkov PhD, L.Ac

Published by :

Nathalie Valkov PhD, L.Ac

4040 Civic Center Dr #200

San Rafael Ca 94903

USA

818-230-2419

nvalkov@ttphc.com

ISBN1450560423

EAN-139781450560429

**Cataloging-in-publication data**

Valkov, Nathalie

  Cordyceps: treating diabetes, cancer and other illnesses/ Nathalie Valkov

  Includes bibliographical references.

  ISBN1450560423

  1.  Health/Fitness 2. Alternative therapies

DISCLAIMER: The information given in this book has been carefully researched. However, the author and publisher cannot accept responsibility for any errors or omission. They will not be held liable for any injury or damages caused to the reader that may result from the reader's acting upon and using the content of this book. This book is intended to be purely informational and is not meant to diagnose. Under no circumstances does it replace the advice of a trained medical practitioner.

# Foreword

I wrote this book so people could actually see and understand the research that has been done on Cordyceps sinensis. Considered a miracle mushroom by many, I felt it was important to see the scientific basis behind it. For those hard core scientists who will read this book, I have included an extensive bibliography that will allow you to obtain more information on the study of your choice. Should you have any questions or comments please do not hesitate to contact me at nvalkov@ttphc.com.

I have tried to make this book as easy to digest and understand as possible. In each section I attempt to give an overview of the western and eastern medical approach being discussed. I also discuss studies and case studies when available.

As I use Cordyceps in my own practice, I see it work every day, and often times in ways that I have never seen discussed. And as amazing as this mushroom may be, it is not without side effects and should not be used without the advice of a trained oriental medical practitioner. Furthermore all the illnesses covered in this book are of a serious nature. If you suspect you have any of them, do consult a health care provider as soon as possible.

I have always resented people who speak their own jargon and make their work inaccessible to others because they cannot use regular language. I may offend the purist by the wording I have chosen to use, but I am hoping that, in doing so, I will make this book more palatable to many more readers.

DISCLAIMER: The information given in this book has been carefully researched. However, the author and publisher cannot accept responsibility for any errors or omission. They will not be held liable for any injury or damages caused to the reader that may result from the reader's acting upon and using the content of this book. This book is intended to be purely informational and is not meant to diagnose. Under no circumstances does it replace the advice of a trained medical practitioner.

# Table of Contents

# Introduction

- ## What is Cordyceps?

Cordyceps sinensis, also known as the caterpillar fungus, summer-plant or winter worm, is a fungus that grows on the larva of a caterpillar. The vegetative body, or mycelium, is enclosed in the body of the insect from which the fungus develops. The fruiting body that buds from the caterpillar has a dark brown base and a black top. It looks like a twig without the typical cap that mushrooms generally exhibit. These fruiting bodies occasionally branch out and recombine to look like deer antlers.

It is made up of the following bioactive ingredients: Cordycepin (3-deoxyadenosine) and Adenosine. It also contains the following amino acids: Aspartic acid, Threonine, Serine, Glutamic acid, Proline, Glycine, Alanine, Valine, Methionine, Isoleucine, Leucine, Tyrosine, Phenylalanine, Histidine, Lysine and Arginine. Its proximate composition is comprised of about 25% carbohydrate, 30% protein, 9% fat, 6% ash and 10% moisture.

Also referred to as Dong Chong Xia Cao, Cordyceps sinensis has been used in oriental medicine for hundreds of years to treat a variety of medical conditions. It is known to the Chinese as being sweet and warm in nature while it enters the Lung and Kidney meridians.

# - How is Cordyceps cultivated?

Cordyceps grows naturally at altitudes of 9,000 to 16,000 feet in the mountains of China, Nepal, and Tibet. Because of its newly found fame in the west, it is also being cultivated quite successfully. Indeed crop growing of this fungus is now being done on grain or in liquid media without affecting the potency of the herb. Modern processes involving electronic control of the agricultural system ensures the quality of the product being grown.

C H Dong and Y J Yao have determined that the following substances provided the best yield for submerged Cordyceps culture. Indeed Mannitol was the best carbon source, while organic nitrogen was the best nitrogen source. Nitrates such as potassium and sodium nitrates also seemed to increase yield. Aspartic acid was the best source of amino acids. A carbon to nitrogen ratio of 12:1 appeared to be ideal. Calcium, although considered non essential, also contributed to a higher Cordyceps yield along with a complete medium of trace-elements. Biotin was the most adapted vitamin for Cordyceps growth. However, when looking at product availability and cost of production, the research team concluded that the best medium for mass production might be sucrose, peptone and yeast extract.

In the past Chinese farmers used a silkworm based medium to grow Cordyceps sinensis. However, organic purple corn has now been determined to be the better option and is thus being used for the growth of this fungus.

## - Health benefits of Cordyceps

There are many health benefits attributed to Cordyceps sinensis. It has been known to the Oriental medical practitioner to tonify the kidneys and the lungs, to stop bleeding to dissolve phlegm and to strengthen Jing (vital energy).

In recent years westerners have seen the benefits of Cordyceps in immune related illnesses such as diabetes, cancer, AIDS and lupus. They have observed kidney function being improved in both acute and chronic renal failure, and liver diseases such as fatty liver, cirrhosis and hepatitis show marked improvement after treatment with Cordyceps. The same is true for lung function issues such as bronchitis, COPD and asthma. Cardiovascular diseases such as arrhythmias, pulmonary heart disease, chronic heart failure and atherosclerosis have also been helped with this herb. Cordyceps is now also known for facilitating the treatment of fatigue and sexual dysfunctions.

## - Guidelines for using Cordyceps and appropriate dosages

Effects can generally start being seen after one week of treatment. However it may take as long as three to six weeks to see more noticeable outcomes.

Capsule forms are usually the easiest to handle. They are simple to use as there is no cooking or mixing required before ingestion. However, Cordyceps can also be found in powder or liquid form, not to mention the actual raw herb. All can be quite effective, but the practitioner must make sure that the Cordyceps used for treatment is standardized and that it comes from a reputable company.

Dosage varies depending on illness and form of the herb used. For instance, when making a tea from the raw herb, 3 to 9 grams of Cordyceps are added to about 4 cups of water and simmered for about 45 minutes until about 2 cups of liquid remains. The patient then drinks about a cup a day. It is the opinion of this practitioner that this method makes tracking the exact amount of herb taken by the patient almost impossible. It therefore seems preferable to use capsules, powders, or liquid extract forms of the fungus where measured quantities can be used.

Dosage will generally be estimated at about three grams of Cordyceps per day. The practitioner should however be aware of the manufacturer's instructions and recommended dosage as fillers may be added in amounts unknown to the practitioner. As with any other herb, self medication is not advised, and seeking the help of a knowledgeable practitioner is strongly suggested.

# 1

## The Quick and Dirty

## on Oriental Medicine

Note: Oriental medical practitioners and people with a basic understanding of oriental medicine should skip this chapter as it is only an extremely basic introduction to oriental medicine.

As Newtonian physics is to Einsteinian physic, western medicine is to oriental medicine. All models are valid in their own rights but use different sets of parameters to establish their theories. That being said, it is important that the westerner gain a basic understanding of oriental medicine in order to fully appreciate what this system can bring.

The theory of yin and yang is the most basic theory that rules oriental medicine. In a nutshell, it states that the nature of Yin is female, passive, negative, and north while yang is male, positive, active and south. They are opposites but also interdependent as one cannot exist without the other (i.e. no day without night). There is also mutual consumption of yin and yang. When all is well, harmony is present. But when there is an imbalance there can be an excess of Yin, an excess of yang, a deficiency (weakness) of yin or a deficiency of yang. Finally there is to possible inter-transformation of yin and yang when one can turn into the other in an orderly manner as is the case with the change in seasons.

The oriental doctor uses a set of organ systems to which functions have been assigned. The liver is responsible for the smooth flow of Qi (or energy) throughout the body and is associated with the gallbladder which stores bile. The heart pumps the blood and also transforms the energy obtained from food into blood. Unlike the western heart, this organ is also associated with emotional balance. The pericardium protects the heart and helps it with its various functions. The spleen takes ingested substances, makes them into Qi and blood, and delivers them to all parts of the body. It is associated with the stomach which is responsible for digestion and the formation of Qi and fluids. Lungs are responsible for the appropriate movement of Qi toward the lower part of the body, the scattering of Qi throughout the whole body and the release of waste products through breathing out. They are associated with the large intestine which transports stool and reabsorbs fluids.

The kidneys are essential in growth, reproduction and development. They also store Essence or Jing, a person's vital energy. They are associated with the urinary bladder which stores and releases urine. They last organ, the San Jiao, is absolutely unknown to the west. It is divided into the upper jiao which contains the heart and lungs, the middle jiao which is comprised of the spleen and stomach, and the lower jiao which is made up of the kidneys and urinary bladder. It is associated with the proper water movement throughout the body.

These organs can be affected by different external and internal influences that will cause diseases. The external factors are the wind, cold, summer heat, dampness, dryness, and fire. The internal ones are sadness, joy, worry, anger, fear, anxiety and terror. It is important to the oriental medical practitioner to maintain the organ mentioned above in perfect condition so that they will not be receptive to possible attacks from the various internal or external causes.

In order to attain this goal the oriental doctor has quite a few options. Acupuncture is probably the most known to the westerner. But he can also use herbs such as Cordyceps sinensis, the one discussed in this book. Other forms of treatment such as moxibustion (the heating of acupuncture points using mugwort) or cupping are also at his or her disposal. They are very effective methods but beyond the scope of this book.

# Part A

# Cordyceps

# And

# Immunity

- Overview of the immune system

Unlike the other body systems, the immune system is a functional one that is comprised of specialized cells and molecules that defend the body against disease.

To produce an immune response the human body uses both the cardiovascular and the lymphatic system.

The cardiovascular system, consisting of the heart, arteries, veins, capillaries and venules is responsible for transporting blood and its constituents to all parts of the body.

The lymphatic system, consisting of lymphatic vessels, lymph nodes, thymus and spleen, carries lymph, a clear watery liquid that contains immune cells, to the interstitial spaces of the body and back into the blood stream.

For an appropriate immune response to occur, the immune system must be able to recognize, amplify and respond to a specific antigen using the immune cells at its disposal without losing its ability to regulate itself.

Antigens, also called immunogens, are foreign substances that produce an immune response. They include bacteria, fungi, viruses, protozoa, and parasitic worms. Because they are made up macromolecules such as proteins and polysaccharides they can be found anywhere. Pollen, poison ivy, and insect bites are among the most common immune response triggers we encounter.

The immune cells responsible for defending the human body are primarily the lymphocytes. B lymphocytes are essential for humoral (antibody triggered) immune responses and mature in the bone marrow. An antibody is a protein present in the blood that finds and destroys antigens by attaching to them. On the other hand, T lymphocytes, which mature in the thymus, are responsible for cell mediated immune responses and also help with the creation of antibodies. According to Carol Mattson Porth T cells make up about 60 to 70% of blood lymphocytes, while B cells make up only 10 to 20%.

At the onset of an immune response an antigen presenting cell such as a macrophage will process and present the antigen to T helper (CD4) cells with the help of an MHCII molecule. When the infection is caused by a virus the infected cell will present the antigen with the help of an MHCI molecule to a T cytotoxic (CD8) cell. During the response T helper cells will release cytokines that will in turn cause B cells to divide and form effector cells that will destroy the antigens via antibody production (passive immunity), and memory cells that provide the body with its ability to recognize the pathogen as foreign as needed (acquired immunity).

It is worth mentioning that we are all born with an innate immunity that is made up of a cough reflex that allows us to cough up foreign particles when they are trapped in the mucus that we produce, enzymes in tears, oils on our skin as well as skin itself and stomach acids which contribute to killing many antigens before they reach the rest of the body.

Inflammation is yet another response linked to our immunity. Without this reaction from our vascularized tissue to specific stimuli our wounds would not heal and minor infections would quickly become overwhelming. During the early stage of inflammation, fluid leaves the capillaries in response to the release of chemicals such as cytokines, thus causing increased blood viscosity. Leukocytes then move along the wall of the capillary and into the area where injury has occurred. Neutrophils and other phagocytic cells arrive first followed by monocytes which become macrophages once at the injury site. They then proceed to engulf the antigen and kill it intracellularly.

With such a complicated network of steps involved in an immune response it is easy to see that many things can go wrong on the many levels involved in an appropriate response and that thus illness can easily develop.

# 2

## Cordyceps and Diabetes

## A.    Disease mechanism

-     Western approach

There are two types of diabetes, Type 1 or insulin dependent and Type 2 or non insulin dependent. The first one is definitely an autoimmune disorder while the second may or may not. Since both types produce the same outcome they will both be discussed together.

Type 1 is caused by an autoimmune destruction of the pancreatic Beta cells that produce insulin and thus leads to a complete lack of the latter with an uncontrolled elevation of blood sugar levels. Islet cell autoantibodies (ICA) and insulin autoantibodies (IAA) appear to be responsible for the development of this disease. A genetic predisposition has also been suggested along with an environmental trigger to stimulate an immune response. MHC genes and genes present on chromosomes 6 and 11 have so far been associated with this disease.

Type 2 is characterized by an increase in fasting blood sugar levels in spite of insulin availability. This state may be achieved because of impaired insulin secretion, insulin resistance, or an increase in glucose production by the liver. The lack of insulin that may arise from a type 2 diabetes is caused by the failure of Beta cells.

The mechanism of type 2 diabetes in obese individuals is yet slightly different. The excess free fatty acids (FFA) stimulate Beta cells to produce more insulin. At the same time the inhibit glucose uptake by peripheral tissues thus preventing glycogen storage in the muscles. Furthermore the increased insulin level caused by the excessive stimulation of the Beta cells will also trigger a release of glucose by the liver. This whole process will eventually cause Beta cell exhaustion followed by an inability to produce insulin.

## - TCM approach

In Chinese medicine Diabetes Mellitus is called "thirsting and wasting" or "Xiao Ke" syndrome. While Type I diabetes appears to be caused by congenital deficiencies, Type II generally starts by stomach heat combined with spleen deficiency.

Heat, when in excess, may be generated from a life-style involving over-eating hot, spicy, or greasy foods along with the ingestion of alcohol. It may also be deficient and come from the liver or stomach. The spleen on the other hand may become deficient with the over-consumption of sugar, sweet or fatty foods, excessive thinking, lack of exercise, or over-exertion.

Kidney deficiency of both Yin and Yang may also evolve from both heat and spleen deficiency. When heat remains present for some time stomach and lung fluids become depleted and may eventually reach the kidney.

Lung dryness may also develop.

Diabetes is also often accompanied by liver Qi stagnation and blood deficiency which in turn develop into to blood stasis. Damp-heat may also arise from spleen Qi deficiency to give rise to some of the complications associated with diabetes.

## B.    Effects of Cordyceps on diabetes

In recent years Cordyceps has been shown to lower blood sugar levels.  Because relatively few studies have been done to date the mechanism by which Cordyceps exerts its effect remains unproven. On the other hand this fungus has been shown to not only lower blood glucose but also increase serum insulin thus helping diabetic patients regulate their blood sugar levels.

It has been suggested by Georges M Halpern MD, PhD, and Andrew H Miller that Cordyceps may affects diabetic patients by calming the immune system cells and thus possibly helping with regulating the autoimmune response. This hypothesis would be a plausible one for type 1 diabetes. However the fungus also lowers blood sugar in type 2 diabetes patients and thus this suggested mechanism cannot be the only one at work in this instance.

In terms of Chinese medicine Cordyceps sinensis will help replenish the kidneys, regenerate fluids, and moisten the lungs. As such, by allowing for the proper distribution of body fluids and Qi, blood stasis may be averted as well.

The appropriate dosage of Cordyceps for this disease should be about 3 to 3.6g per day in order to observe marked changes.

## C.    Research studies on the subject

In "Hypoglycemic activity of polysaccharide, with antioxidation, isolated from cultured Cordyceps mycelia" the authors described how the polysaccharide CSP-1 has had a hypoglycemic effect on STZ-induced diabetic rats and alloxen-induced diabetic mice while it did not significantly affect normal mice. A drop in blood sugar levels was observed along with an increase in serum insulin levels in the diabetic animals when a dose higher than 200mg/kg of body weight was given daily for 7 days.

Conclusion: The authors concluded that CSP-1 possibly stimulates the pancreas to release more insulin and/or that CSP-1 may reduce insulin metabolism.

In "The anti-hyperglycemic activity of the fruiting body of Cordyceps in diabetic rats induced by nicotinamide and streptozotocin" the authors showed that the fruiting body of Cordyceps reduced the diabetes induced weight loss, polydipsia and hyperglycemia in animals suffering from diabetes. They further suggested that because of these findings the fruiting body of this fungus may be incorporated in the diet of diabetes sufferers to naturally reduce their blood sugar levels and related symptoms. The setup of the experiment was as follows. An STZ group was administered with placebo, another (FB) was given 1g of fruiting body per day, another (CC) was given 1g of carcass, and another group (CF) was given 1g of fruiting body plus carcass for four week while all groups were being injected with nicotinamide and streptozotocin. A control group (CON) was fed with placebo and given saline injections. The results showed an increase in water consumption, fasting blood glucose concentration, and serum fructosamine concentration in the STZ, CC and CF groups. FB and CON groups showed much lower increases thus proving the effectiveness of the Cordyceps fruiting body in controlling diabetes.

Conclusion: The fruiting body of Cordyceps sinensis is effective in controlling diabetes.

In a further study described in "Anti-hyperglycemic Activity of Natural and Fermented Cordyceps sinensis in Rats with Diabetes induced by Nicotinamide and Steptozotocin" the authors conclude that the use of the fruiting body of the fungus, the fermented mycelia or its broth offer similar clinical results and should therefore also be considered for the treatment of diabetes. The experiment was set up as follows. . An STZ group was administered with placebo, another (FB) was given 1g of fruiting body per day, another (MCS) 1g of fermented mycelia, another (BCS) 1g of fermented broth of Cordyceps, and another group (XCS) was given 0.5g of fermented mycelia and 0.5g of fermented broth. Each group was injected with nicotinamide and streptozotocin. A control group (CON) was injected with saline and fed placebo. The results were as follows: for groups FB, MCS, BCS and XCS water and food consumption, blood glucose level two hours after eating and fructosamine serum level were much lower than with the STZ group

Conclusion: Both broth and fermented mycelia are effective in the treatment of diabetes.

In August 2000 researchers from the Jinke Medical Institution published a study on the effect of Cordyceps sinensis in the treatment of diabetes combined with hyperlipidemia. Subjects were given 4 capsules of a combination of herbs including Cordyceps sinensis, 3 times a day, and 5 minutes before meals for a period of 6 months. 48.3% of the test subjects showed marked improvement, and 41.9% showed improvement, while only 9.7% showed no improvement.

Conclusion: Cordyceps sinensis is effective in the treatment of diabetes.

These studies appear to indicate that Cordyceps indeed improves blood sugar levels in individuals treated with this fungus. The results on rats are undeniable and the applications to human subjects seem more than promising. Furthermore, as Cordyceps sinensis is being used in greater quantities worldwide, it is particularly important to be able to use different preparations of the fungus with proven similar results.

# D. Case studies

A 65 year old diabetic patient had had erratic blood sugar levels for 3 years. Her numbers varied from 300 to 180g/dl. She was then placed on Cordyceps (1.8g twice daily). Within one week her blood sugar levels were consistently below 180g/dl. Because her blood sugar levels tended to be higher in the morning the evening dosage of Cordyceps was increased to to 2.4g while the morning dosage remained the same. Within one month her blood sugar levels were completely regulated and consistently under 120g/dl.

A 64 year old man had been on insulin (15 units per day) for 6 months after his western medication had stopped producing the desired effect. Due to unforeseen circumstances during a trip the patient had to stop taking his insulin. When he came back from his trip his blood sugar levels were consistently over 300g/dl. He was then started on a treatment of 1.8g of Cordyceps twice daily. No marked improvements were seen the first week. However a noticeable change was seen in his blood sugar levels at the end of a month. Since morning blood sugar levels were still elevated, the evening dosage of Cordyceps was increased to 2.4g. Within one week after that his blood sugar levels were consistently below 140g/dl without the help of insulin. The patient has now been insulin free for over one year and all blood tests performed in the lab appear normal. As an added bonus, this patient, who also happens to be my father, was able to decrease his metformin and glyburide (other diabetes drugs) dosage by half by just adding a few other herbs to the Cordyceps.

# 3

## Cordyceps and Cancer

## A. <u>Disease mechanism</u>

-    Western approach

Cancer is an immune related illness in which the patient's body has lost its ability to recognize a tumor cell as foreign thus allowing the formation of malignant growths or tumors. Tumor cells can also travel through the body undetected and colonize various organ systems.

Cancer stems from an altered cell differentiation and growth that has become unregulated. It can start in any organ or body system and affects men and women almost equally.

A normal cell cycle involves four phases during which DNA, RNA and protein synthesis occur so that the cell can separate into two identical cells.

All the cells in the human body come from the zygote created at conception. Based on the environment the cell develops in, such as the external stimuli provided by nearby cells, nutrients, oxygen and ions, a cell will become much more specialized. The more specialized they are the less they are able to divide. It is thus believed that cancer cells develop from normal cells that have undergone a mutation during the differentiation process. If the mutation happens early on, the resulting cell will be very poorly differentiated and the cancer will be very aggressive. However if the mutation happens at a later stage the cells will be more differentiated and will thus produce a much less aggressive cancer.

Any number of environmental factors can affect the differentiation process. Radiation or toxins such as pollutants can cause cellular damage.

The many different types of cells present in the human body and the complexity of the human body itself give rise to many different types of cancers possible. But in spite of such diversity, every cancer is due to the lack of the host body to recognize the cancer cell as a harmful antigen. A treatment stimulating the host's immune system to trigger a more aggressive response against cancer cells may be one way tame the disease.

- TCM approach

On the other hand Chinese medicine considers that external pathogenic factors invading the body, along with dietary and emotional causes, will affect the proper functioning of the Zang-Fu organs and hinder the circulation of Qi and Blood thus causing Qi stagnation and Blood stasis as well as the accumulation and stasis of Phlegm-damp. This buildup will in turn lead to the formation of tumors.

## B. Effects of Cordyceps on cancer

In recent years Cordyceps has been shown to help cancer patients. Those suffering from lung cancer seem to reap the most benefit from this type of therapy even if it has been shown to help cancers affecting the cervix, vagina, prostate, breast, liver, gastrointestinal tract, pancreas, lymph nodes, larynx and smooth muscles. Cordyceps indeed seems to increase T cells, and lymphocyte production in general.

General symptoms associated with cancer appear to improve greatly when the treatment involves Cordyceps. Fever is reduced. Patients feel hungrier and experience less vomiting, fatigue and pain. Complete Blood Count (CBC) also remains normal in more patients undergoing chemotherapy than in patients who do not take the fungus. Cancer patients on Cordyceps have also been known to complete their chemotherapy in higher numbers thus having a better chance of survival.

Cordyceps has also been shown to reduce the growth of cancer cells in cases of lymphomas, melanomas and sarcomas. It also enhances the effect of some chemotherapies.

The exact mechanism by which Cordyceps helps fight cancer is still relatively unproven. Its ability to enhance immunity is well documented but the means by which it seems to improve the human body ability to trigger an immune response against cancer cells is still not understood.

## C.    Research studies on the subject

In 1993, Liu Feng and Zheng Xiao published an article on the study of Cordyceps sinensis and its effect on Laryngeal carcinoma. In their research they used laryngeal carcinoma cells in vitro with a fetal calf serum based culture solution along with different concentrations of Cordyceps sinensis water extract. They proceeded to count the number of cells present at different time intervals and for the different concentrations. Their results not only showed that the less diluted the extract was the more potent the fungus appeared to be on laryngeal carcinoma cells, but that the extract was also time sensitive. Indeed the longer the cells were exposed to the fungus the slower they grew. About $18 \times 10^{24}$ cells were alive at 24 hours while only about $11 \times 10^{24}$ cells were alive at 120 hours with the 1:50 dilution. With lesser dilutions there results were not as drastic but did show a growth rate reduction as well.

Conclusion: It was thus the authors' conclusion that Cordyceps sinensis has a direct impact on laryngeal carcinoma cells as it was noted that growth rate was reduced with higher concentrations of Cordyceps and over time. They suggested that further studied should be made to determine the exact mechanism by which these results were obtained.

In 1998, Kazuki Nakamura and his team published a study on the inhibitory effect of Cordyceps sinensis on spontaneous liver metastasis of Lewis lung carcinoma and B16 melanoma cells in mice. The C57BL/6 mice were injected with Lewis lung carcinoma cells and were killed 20 days after inoculation while the mice injected with B16 melanoma cells were then killed after 26 days. They found that the water extract of Cordyceps sinensis had a strong inhibitory effect on both types of cells while cordycepin alone did not. The actual results for the water extract of Cordyceps sinensis were as follows. The average weight of the primary tumor in the control group was 3.49g while the weight of the primary tumor after ingestion of 100mg/kg of body weight was 2.80g for Lewis lung carcinoma while the average weight of the primary tumor in the control group was 2.21g while the weight of the primary tumor after ingestion of 100mg/kg of body weight was 2.81g and 1.17g after ingestion of 200mg/kg of body weight.

Conclusion: The water extract of Cordyceps sinensis had a strong inhibitory effect on both types of cancer cells.

Kazuki Nakamura and his team studied the activation of in vivo Kupffer cell (liver macrophages) function by oral administration of Cordyceps sinensis in rats in an attempt to find the mechanism explaining the anti-metastatic properties of Cordyceps sinensis. The animals were given water extract of Cordyceps sinensis at a dose of 200mg/kg of body weight for 25 days and were then injected with a colloidal carbon solution. The rate of carbon clearance from the blood was then measured along with body, liver and spleen weight. The experiment results showed that the half-life of the colloidal carbon in the blood of the rats who had received the water extract of

Cordyceps sinensis was much shorter (36%) than their control counterparts. The body, liver and spleen weights remained constant.

Conclusion: The Kupffer cells of the animals treated with the fungus had accelerated function and that because they are responsible for scavenging tumor cells they were therefore partially implicated in the anti-metastatic function of Cordyceps sinensis. Furthermore the lack in weight change showed that the use of Cordyceps had no toxic effects on the body and that they were thus quite safe to use.

In 1997 Yu-Jen Chen and his team studied the effect of Cordyceps sinensis on human leukemic U937 cells in vitro. Some cells were treated with a polysaccharide fraction extracted from Cordyceps sinensis while others were treated with a conditioned media containing mononuclear cells from human peripheral blood as well as the Cordyceps extract. Control group of untreated cells and cells only treated with mononuclear cells were also setup. Results showed that cell proliferation was inhibited by the mononuclear cells/Cordyceps extract while the controls and Cordyceps extract alone were not. Inhibition rate for the mononuclear cell/Cordyceps mixture was indeed observed to be over 83% after day 5. The number of undifferentiated cells also decreased from over 95% to about 27% with this mixture while the number of monocytes increased from 0% to about 50%.

Conclusion: The authors concluded that because the Cordyceps alone could not produce significant results while the monuclear cell/Cordyceps could, the antitumor mechanism must involve the activation of the host immune response. Indeed mononuclear cells were induced into producing higher cytokine levels that do have anti proliferative and differentiation- inductive properties on leukemic U937 cells.

In 2007, Jian Yong Wu, Qiao Xia Zhang and Po Hong Leung studied the inhibitory effects of ethyl acetate extract of Cordyceps sinensis mycelium on various cancer cells in culture and B16 melanoma in mice. Cordyceps extracts were obtained with petroleum ether, Ethyl acetate, ethanol and hot water. Results were obtained for breast cancer cells (MCF-7), mouse melanoma (B16), human premyelocytic leukemia cells (HL-60), and human hepatocellular carcinoma (HepG2). Tests were performed in vitro on cultures of the cancer cells plated on fetal bovine serum and antibiotics, and in vivo on C57BL/6 male mice for mouse melanoma. The results indicated a strong anti-tumor activity of all Cordyceps extract except with hot water. More specifically the ethyl acetate Cordyceps extract caused a 60% decrease in tumor size over the 27 days treatment of the B16 melanoma in vivo.

Conclusion: Cordyceps sinensis showed strong antitumor activity.

Over the years the effectiveness of Cordyceps on various types of cancers has been demonstrated over and over. From breast cancer, to leukemia, to melanoma, to lung and liver cancers, the use of Cordyceps appears essential for the survival of the patient being treated. These studies appear to be a confirmation of what popular literature has been claiming now for the last few years.

## D.   Case study

A 66 year old woman suffered from lung cancer affecting her right lobe. Between chemotherapy which produced no result and radiation therapy which eventually killed her because of radiation induced pneumonia, the patient was treated for 3 weeks with a Chinese herbal formula containing Cordyceps sinensis. During that time the tumor went from a diameter of 7cm to 6 cm. Once radiation started, the tumor did not grow nor shrink until the patient's death due to complications associated with radiation. No metastases were ever developed.

My own grand-mother was diagnosed with breast cancer over a year ago. She has opted, against everyone's advice (including mine), to only treat her cancer with herbs.

Cordyceps was of course among the herbs I prescribed for her. I am happy to report that as of today, her tumor has shrunk and no metastases have been detected. I would not tell anyone to follow this path in treating their cancer, but I must admit that it worked for her and that her quality of life has been maintained. Always seek professional help when dealing with illnesses as severe as cancer. In my opinion, a combined treatment allowing both western and oriental medicine to work together is the best approach.

# 4

## Cordyceps and AIDS

## A.   Disease mechanism

-    Western approach

According to the Merck Manual, AIDS or Acquired Immunodeficiency Syndrome is a secondary immunodeficiency syndrome caused by a prior HIV infection and characterized by opportunistic infections, cancers, neurologic dysfunctions and various other diseases.

This retrovirus (HIV or Human Immunodeficiency Virus) invades the body by sexual contact, blood contact, or perinatally. Like all other retroviruses it carries its genetic information in its RNA. When it is transmitted to a new host, it infects CD4 T cells specifically by presenting a surface protein called gp120 to the target cell and binding to its surface. The virus then enters the hosts cell and dumps its viral RNA to be transcribed into DNA with the help of the reverse transcriptase enzyme. The resulting DNA is then integrated into the host's DNA. From that point on the virus may either lay dormant or replicate itself and release the copies, destroying the host's cell in the process.

In a given day, almost all CD4 T cells are replaced and almost all the viruses are destroyed. But over time the CD4 T cell count decreases while the number of viruses increases. The host's immunity thus decreases as well, as the CD4 T cells play such an important role in the immune response. The host then becomes more prone to infections and cancer.

Once the host has less than 200 CD4 T cells remaining per cubic millimeter of blood and 26 clinical presentations such as infections, the host is now said to have AIDS.

- TCM approach

According to traditional Chinese Medicine excessive and unprotected sex will lead to the exhaustion of kidney Jing which in turn will make the individual more susceptible to toxic pathogens. Drugs, which tend to be drying and hot in nature, will deplete yin and body fluids thus causing yin deficiency. The toxic pathogen which may be dormant for a while may become activated when another seasonal pathogen attacks the carrier. The toxic pathogen may also enter the blood level directly when the future host receives a tainted blood transfusion.

According to Anshen Shi, HIV and AIDS stages can be expressed in terms of the four levels of warm febrile diseases. The strength of both the defensive Qi and the toxic pathogen will determine the speed at which the illness develops. At onset of infection a deficient defensive Qi and an unrelenting toxic pathogen are present. The pathogen is attacking the lung and stomach yin. As the defensive Qi weakens the toxic pathogen moves in deeper into the nutritive and blood levels. The Qi is sinking and heat accumulates in the stomach and intestines. The defensive Qi becomes too weak to expel the pathogen. Phlegm heat is misting the orifices and becomes lodged in the terminal yin. Eventually the toxic pathogen takes over completely. At this point Yin and Yang separate and the prognosis is very poor.

## B.  Effects of Cordyceps on HIV/AIDS

According to Jian Wang and Wen Zou, Cordyceps appears to help in the treatment of HIV/AIDS by enhancing the CD4 helper T cells and promoting the CD4/CD8 ratio. It promotes the proliferation of T cells as well as strengthening their function. So overall it improves the immune system.

Furthermore Cordyceps is known to tonify Yang, strengthen lung yin and transform phlegm and also stop bleeding. In the early stages of the disease Cordyceps can be used to strengthen lung yin. As the disease progresses it can be used to transform phlegm and finally to tonify Yang. As symptoms of bloody diarrhea are common, Cordyceps may be able to help with resolving these issues as well.

## C.  Research studies on the subject

In 1989 David C. Montefiori and co-researchers studied the effect of phosphorothioate and cordycepin analogues of 2', 5'-oligoadenylate on the inhibition of Human Immunodeficiency Virus type 1 reverse transcriptase and infection in vitro. Cordycepin analogues demonstrated up to about 70% inhibition of reverse transcriptase activity while showing having an anti-HIV1 activity leading to over 80% protection. The team determined that even though complete inhibition of reverse transcriptase was never achieved, a strong antiviral activity was still observed. They thus concluded that a mechanism other than reverse transcriptase inhibition may be responsible for the antiviral properties observed. They further suggested that more research should be done so determine the precise mechanism (i.e. RNase L activation or other unknown mechanism) by which anti-HIV1 activity can be explained.

In 1995, Lu Weibo and colleagues published a research on the effect of traditional Chinese medicine on HIV and AIDS patients. Among the herbs dispensed was Cordyceps sinensis. 8 patients from Tanzania were sero-negative converted after their herbal treatment. They were then tested over the years to see if they because sero-positive again. During their initial treatment, improvement in immune function was observed in all cases until they sero-negative converted. Of all those cases only one of the patients returned sero-positive status after 3 months of sero-negativity. The sero-negative status of the others appeared to last between 11 and 49 months, while 2 of the patients remained sero-negative until the end of the study.

Conclusion: Cordyceps sinensis showed strong antiviral activity sufficient to at least temporarily reverse the sero positive status of the patients.

The use of Cordyceps sinensis on AIDS/HIV patients appears to indeed improve their chances of survival by stimulating an immune response.

## D.   Case study

This case study was reported in "a report on 8 sero-negative converted HIV/AIDS patients with traditional Chinese medicine".

"Patient 5, male, 20 years in age, single, complained of asthenia, dyspnea, pruritus, throat distress and productive cough for 2 months. The Karnovsky Score was 80 points. Lab findings: ELISA (+). His mother was also ELISA (+) too. Then 809 was given to the patient. The asthenia disappeared immediately. 87 days after the treatment, the ELISA was converted to negative. $T_4CC$ was $659.71/mm^3$ and $T_4 /T_8$ ratio reached 1.60. In August 1990 and February 1993 the patient was rechecked 6 times. The results were: ELISA (-), ELISA (+) but WB was indeterminate. In April 1991, both ELISA and WB were negative again with 3 successive ELISA (-) in examination over 34 months; while

$T_4CC$ was 509.37 and 559.67, $T_4/T_8$ ratio was 1.89 and 1.78 respectively. During the treatment, the skin rash and cough subsided gradually, while the body weight was stable. Till September 1991, the patient was free from any complaint. PCR was done in August 1992. It was positive and the p24 antigen was negative. One year later, a second PCR with 2 pairs of primers was repeated and it showed the same result as in 1992."

# 5

## Cordyceps and Lupus

### A.   Disease mechanism

-   Western approach

Systemic lupus erythematosus or SLE is another disease of the immune system where chronic inflammation of the connective tissues is exhibited during periods of exacerbation. The onset may be rapid following a fever or may take years to develop. Western medicine considers its cause to be unknown. in any case, it is caused by an autoimmune response where autoantibodies and immune complexes are formed.

 A B-cell hyperactivity can be detected in patients suffering from SLE. This hyperactivity which leads to overproduction of antibodies may result from overzealous helper T cells or faulty suppressor T cell functions.

Environmental, immunologic, hormonal and/or genetic factors may be responsible for the development of autoantibodies. Ultraviolet light, chemical, certain foods and possibly infectious antigens may cause the onset of the disease and the following exacerbation periods. Patients suffering from SLE may become photosensitive.

 The immunologic component lies in the possibility that the disease process starts with the activation of polyclonal B cells that produce excessive amounts of autoantibodies. These autoantibodies then attach to the appropriate antigen to make immune complexes. These complexes are then dumped into vascular and tissue surfaces where they start an inflammatory response and thus damage the surrounding tissue. Antibodies associated with SLE include ANA or antinuclear antibodies. According to Carol Mattson Porth other types may be produced to attach to red blood cells surface antigens, platelets, and coagulation factors thus leading to anemia and/or thrombocytopenia.

Androgens seem to protect against SLE while estrogen seem to support the development of the disease. An imbalance in estrogen/androgen levels may cause the increase in the helper T cells activity while reducing the suppressor T cell response thus leading to the formation of autoantibodies.

Familial cases especially among identical twins appear to substantiate the idea that development of autoantibodies may have a genetic cause. Furthermore four genes associated with the disease have been isolated so far and thus corroborate the genetic component theory.

-    TCM approach

On the other hand Traditional Chinese medicine believes that SLE is associated to an excessive heat toxicity and a yin and fluid deficiency.

Yin deficiency can be either be constitutional or caused by a variety of different pathogens. This deficiency may then lead to an imbalance between the nutritive and defensive Qi, which in turn will increase the chances of a successful external attack.

According to Anshen Shi, wind, cold, summer heat, dampness, dryness and heat can all attack and invade the body thus causing SLE. Heat toxicity may attack the blood vessels and bring on erythema between the skin and muscles, or joint swelling when it happens between the joints. It may also attack the internal organs thus causing organs specific signs and symptoms.

SLE may also be caused by emotional stresses, over-eating, excessive alcohol consumption, over-exertion and excessive sexual activity. But because it often starts after exposure to sunlight, traditional Chinese medicine believes that the six external pathogens mentioned above are the main causes of the disease.

SLE goes from excess in its early stage to deficiency latter on. In the early stage heat will injure yin. Yin will become deficient. Yin deficiency with empty fire, the pattern associated with the chronic active stage of the disease, will then deplete yin and body fluid. Finally Qi and yin, or yin and yang will both be deficient during the late stage of SLE.

At the same time SLE can be seen as moving from the exterior (superficial layers such as skin and mucosa) to the interior (internal organs).

## B.    Effects of Cordyceps on lupus

Cordyceps sinensis appears to prolong the life span of patients suffering from SLE. It has been shown to inhibit anti-ds DNA where the antibody production can be completely stopped in many cases. The effects of this fungus on the quality of life of these patients seems to be an added bonus as people on a Cordyceps regimen report have more energy and generally less stress and fatigue.

## C.    Research studies on the subject

In 1993, Jong-Rern Chen and his team studied the effects of Chinese herbs (including Cordyceps sinensis) on improving survival and inhibiting anti-ds DNA antibody production in Lupus mice. During this experiment their used NZB/NZW F1 mice that were fed 0.025, 0.05 or 0.1 g/day of Cordyceps solution. The results indicated that the group fed with the 0.1 g/day solution prolonged the life span of the mice and inhibited anti-ds DNA production. The control showed a survival rate of 0% after 8 months of observation while the Cordyceps survival rate was at 75%. Similarly about six months into the experiment the negative percentage of anti-ds DNA was 43% for the control group with 14% dead mice while the Cordyceps group had an 87.5% negative anti-ds DNA rate and 0% dead mice, thus showing clearly that Cordyceps sinensis has a definite impact on SLE.

Conclusion: Cordyceps sinensis has a definite impact on SLE

In 2001 researchers from Zhujiang Hospital published a study on the effect of Cordyceps sinensis on inhibiting SLE in rats. The subjects were MRL 1pr/pr female rats that were fed ethanol dissolved Cordyceps sinensis at the rate of 40μg/kg/d for 8 weeks. The control group was just fed ethanol. Lymph node assessment showed better results for the Cordyceps group. Proteinurea showed a marked decrease in the Cordyceps group while the control showed a net increase. Serum urea nitrogen experienced the same trend while creatinine levels showed a slow decrease for the Cordyceps group and a sharp increase for the control group. Finally ds-DNA antibody levels decreased from about 0.16 to about 0.14 for the Cordyceps group while the control group went from about 0.15 to about 0.2.

Conclusion: The researchers concluded that Cordyceps sinensis is indeed effective in inhibiting lymphadenectasis, reducing proteinurea levels and ds-DNA antibody while improving kidney function, all signs of SLE.

In 2003 L. Lu and his colleagues studied the effect of Cordyceps sinensis and artemisinin in preventing recurrence of lupus nephritis. In the research 61 SLE sufferers were treated. 31 individuals received 2 to 4g of Cordyceps powder and 0.6g of artemisinin per day for 3 years. The control group was given tripterygiitotorum and/or

Baoshenkang tablets. The results showed that the Cordyceps mixture was very effective in 83.9% of the cases, effective in 12.9% of the time, and ineffective 3.2% of the time, while the control; group showed high effectiveness in 50% of the cases, effectiveness in 26.7% of the cases and ineffectiveness in 23.3% of the cases. Furthermore, the participants suffered less from side effects when treated with the Cordyceps mixture.

Conclusion: It was the author's conclusion that Cordyceps and artemisinin could stop the reappearance of lupus nephritis as well as protect the kidneys.

Overall these studies seem to agree that Cordyceps sinensis could greatly benefit many sufferers of SLE if incorporated into their treatment protocol.

# Part B

# Cordyceps

# And

# Kidney functions

- Overview of western kidney functions

The kidneys, located against the posterior wall of the abdomen between T12 and L3 on either side of the vertebral column, are made up of a cortex and a medulla which is in turn made up of renal pyramids. At the end of each pyramid are the calyces which receive the urine that leaves the renal papilla (a subdivision of the renal pyramids). From there the urine is transported to the renal pelvis which eventually becomes the ureter as it exits the helium (notch on the medial surface of the kidney).

The kidneys are highly vascularized, as 1,200ml of blood go through them every minute. As arteries take blood to the kidneys they branch out to become arterioles which in turn become capillaries. The dense network of capillaries that results is called a glomerulus.

The main function of the kidneys is to filter blood to make urine. This function is essential as it allows for the body's homeostasis to be maintained. Indeed sodium, potassium, chloride and nitrogenous waste such as urea exit the body as needed through the kidneys. The kidneys are also responsible for pH regulation. Indeed $H^+$ ions are excreted in the urine as needed to maintain optimum pH. Furthermore the kidneys secrete large amounts of organic anions obtained from drugs or made by the body itself, as well as organic cations. It is worth mentioning that only drugs that are not bound to plasma proteins will be eliminated in the urine.

The kidneys also affect the output of ADH (anti diuretic hormone) and aldosterone which is responsible regulating sodium and potassium elimination. They also activate Vitamin D, and produce erythropoietin and some prostaglandins. Renin, one of the hormones responsible for regulating blood pressure, is also produced and stored in the kidneys.

A nephron, the basic functional unit of a kidney, is made up of a renal capsule which in turn contains a Bowman's capsule and a proximal convoluted tubule, a Loop of Henle, a distal convoluted tubule and a collecting duct.

Filtration, reabsorption and secretion occur in the nephron. Filtration is the movement of water and solutes from the plasma through the glomerulus and into the Bowman's capsule. Reabsorption is the movement of the molecules back into the blood, while tubular secretion is the movement of the molecules from the blood to the tubules for excretion.

BUN and creatinine levels are generally taken to determine the functionality of the kidneys.

# Overview of oriental kidney functions

According to Giovanni Maciocia, the kidney is considered to store Jing, govern birth, growth, reproduction and development. It also produces marrow, fills up the brain and controls bone. It governs water, controls the reception of Qi, opens into the ears, manifests on the hair, controls the lower orifices, and houses will power.

The essence of the kidney is first inherited from the parents then restocked with the Qi obtained from the food we eat. The inherited part of Jing, also known as Pre-Heaven Essence, is responsible for proper fetal development, for growth and development after birth, puberty, and fertility. The acquired part of Jing, or Post-Heaven Essence, is stored in the kidneys after having been extracted from food and transformed by internal organs. A decline in kidney Jing will lead to aging. Kidney Essence is also the basic for kidney yin and yang which are in turn the basic of all the body's energies.

From kidney Jing is also derived the production of Marrow which refers to the common medium of bones, bone marrow, spinal cord and brain. Marrow produces the spinal cord and nourishes the brain thus allowing for good memory, concentration, and sight. Along with mental strength, the kidneys are also responsible for bone strength.

Furthermore it is the Jing that nourishes the ears and allows for their proper functioning. It is also essential for suitable hair growth, shine and color. Declining kidney Essence will indeed lead to hair graying.

The kidneys are involved with the transformation and transportation of body fluids. They are described as gates that open and close as needed to control the flow of body fluids into the lower Jiao. Furthermore the kidneys supply the bladder with the Qi it needs to store and transform urine. Even the separation of clean from dirty fluids by the intestines is controlled by the same organ. The kidneys are also involved with the lungs in receiving and vaporizing fluids back up to the same organs to keep them humidified, and with the spleen to supply the necessary heat to transform and transport fluids.

The kidney also receives the air Qi from the lungs and keeps it down. At the same time, it is responsible for controlling the lower orifices and thus incontinence and nocturnal emissions.

Even though it is not technically a kidney function, it is worth mentioning the Ming Men, or Gate of Vitality, which resides between the two kidneys and provides heat to the whole body. According to Giovanni Maciocia "The Gate of Vitality is the embodiment of the fire within the Kidneys".

# 6

## Cordyceps and Renal failure

## Chronic renal failure

## A.    Disease mechanism

-    Western approach

According to western medicine chronic renal failure is the progressive and permanent damage of the kidneys that is characterized by reduced glomerular filtration and tubular reabsorption, as well as a decrease in hormone production.

The destruction of nephrons may be caused by hypertension, urinary tract infections or obstructions, congenital kidney defects, glomerular dysfunction, diabetes Mellitus or SLE.

According to Carol Mattson Porth, the four stages associated with chronic renal failure are diminished renal reserve, renal insufficiency, renal failure and end-stage renal disease.

Diminished renal reserve is characterized by a glomerular filtration rate that is 50% of normal while the BUN and creatinine levels remain within normal range. The patient is asymptomatic and there are no overt signs of decreased renal function.

As the glomerular filtration rate decreases from 50% of normal to 20% the disease is said to have progressed to the renal insufficiency stage. Increased nitrogenous wastes, anemia and hypertension generally develop. As the functional nephrons take over the job of the destroyed ones more blood is filtered and more solutes move out. This causes more water to filter out as well as these solute are osmotically active. The resulting symptom is excessive urination with a urine salt concentration similar to that of blood plasma.

When the glomerular filtration rate is below 20 to 25% of normal the patient is now experiencing renal failure. Edema, metabolic acidosis and elevated blood calcium levels result.

When the glomerular filtration rate falls below 5% of normal, the patient is said to suffer from end-stage renal disease. At this stage, dialysis or transplant is essential. Renal mass is decreased, scarring is observed inside the glomeruli and the number of renal capillaries is reduced.

-      TCM approach

Traditional Oriental medicine considers chronic renal failure to be caused by a spleen and kidney deficiency with damp accumulation. As the kidney governs water and the spleen is responsible for transportation and transformation, an increased kidney Qi deficiency will disrupt the spleen's function and thus increase dampness. As the spleen becomes more deficient, dampness not only accumulated but Qi and Blood are not generated and signs of renal failure such as anemia develop.

According to Anshen Shi, exterior attacks, interior injury and improper diet are associated with the development of the disease as well.

When the exterior pathogen attacks the lungs and causes a lung Qi deficiency water metabolism is impaired and dampness accumulation results. This condition further exacerbates the existing deficiency.

Overwork and excessive sex will lead to further internal injury and will thus increase spleen and kidney deficiency.

An improper diet will tax the spleen which will in turn cause more dampness accumulation and thus increase yet again the already present spleen and kidney deficiency.

## B.    Effects of Cordyceps on chronic renal failure

Cordyceps sinensis has been shown to increase and redistribute epidermal growth factors of renal tissue, as well as speeding up the recovery of renal tubules, thus promoting the recovery of renal function.

Research also shows that Cordyceps sinensis helps reduce the influx of calcium into the renal cortex while protecting the activity of the ATP enzyme, reducing peroxidation of cellular lipids, promoting DNA formation, and improving energy metabolism of mitochondria.

## C.    Research studies on the subject

In 1984 Researchers from Shanghai University of Traditional Chinese Medicine published a study on the treatment of chronic renal failure using primarily Cordyceps sinensis. Patients were to take 4.5 to 6 g of Cordyceps per day, decocted and drank with the herb residue. Average values before treatment were 6.98 for Creatinine, 56.56

for BUN, 17.72 the elimination rate of endogenous creatinine (Ccr). After a course of treatment (about 2.6 months) the results were as follows: creatinine was down to 6.09 and BUN to 48.65 while Ccr had increased to 20.87.

Conclusion: These results demonstrated clearly the positive impact that Cordyceps sinensis has on the treatment of chronic renal failure.

In 2008, researchers from the Shandong Qian Fo Shan Hospital published a study on the effect of combining Chinese and western medicine for the treatment of chronic renal failure. 80 subjects were entered in this study. 40 patients received 3 to 5g/day of Cordyceps sinensis orally and a rhubarb-aconite decoction rectally as a retention enema once or twice daily along with the usual non dialysis treatment. The control group only received the western treatment. The results were as follows; from the control group 23 patients improved, 14 are in stable condition, 3 became worse and no obvious decline in BUN and Scr were observed. The group receiving both Chinese and western treatments showed improvement in 37 cases, 2 were stable, 1 aggravated and both BUN and Scr were markedly decreased.

Conclusion: The researchers concluded that even though western medicine provided some improvement, the combination of both western and Chinese medicine was even better.

Recent studies have shown that Cordyceps sinensis can indeed help with the treatment of chronic renal failure.

# D.   Case study

This case study was obtained from "Initial Observation of 28 Cases with Chronic Kidney Failure Treated Mainly with Cordyceps Sinensis".

A 31 year old male was hospitalized with soreness in the waist, slight edema and proteinurea. He had had nephritis 3 years prior. About 2 years ago, his blood pressure rose around 150 to 220/100 to 128. Hypotensors showed to be ineffective. In the last six months he developed symptoms such as nausea, vomiting, fatigue and cramps. Upon hospitalization lab test results were as follows: SGPT 300 units, Creatinine 5.6mg%, urea nitrogen 53.5mg%, Carbon dioxide 38.5vol%, calcium 11.7mg%, phosphorus 7.5mg%, HBsAg 1:256 positive. SGPT decreased to normal levels and he was transferred to the nephritis ward. His lab tests were as follows: Creatinine 7.9mg%, urea nitrogen 51.3mg%, Hemoglobin 7.5mg%, 7.65g fixed amount of proteinurea in 24 hours. Dizziness, nausea, skin itching, cramps, a white greasy thick tongue coat were present. Initially salvia and glucose injections were given but 4 weeks later creatinine levels had risen to 9.75mg%, so 6g of Cordyceps were given instead. After two weeks his creatinine levels had decreased to 7.5mg%, to 3.2mg% after a month while the hemoglobin increased to 9.8g% and proteinurea decreased to 3.45g. Other symptoms disappeared as well, so he was released and returned home to continue treatment.

# Acute renal failure

## A.  Disease mechanism

-  Western approach

Acute renal failure is characterized by an abrupt decrease in kidney function leading to an increase in nitrogenous wastes, electrolyte and fluid imbalance. It is most commonly seen in the elderly and very ill patients already in an intensive care unit. It is generally associated with other diseases such as heart failure or shock.

The accumulation of nitrogenous waste in the blood is due to a decrease in glomerular filtration rate. Little or no urine is produced and thus little or no waste is excreted.

There are three possible types of acute renal failures. They may be prerenal, intrarenal or postrenal.

In prerenal failure, the renal blood flow is somehow impaired. Possible reasons may include severe hemorrhaging or dehydration, cardiogenic shock, anaphylaxis, sepsis, or mechanical blockage of renal blood flow such as surgeries associated with the kidneys. NSAIDs, because of their ability to inhibit prostaglandin (responsible for dilation and constriction in vascular smooth muscle cells) production, may also be responsible for prerenal failure in patients with reduced renal blood flow.

Intrarenal failure occurs when the site of injury is within the kidneys, as is the case with the tubules and glomeruli. Tubular injury may be caused by a lack of oxygen being delivered to the area, toxic agents, or intratubular obstruction.

Postrenal failure is caused by the obstruction of the urine flow from the kidneys anywhere from the ureter to the urethra. Calculi, strictures, tumors, neurogenic bladder, prostatic hypertrophy or hyperplasia may all be causes for the obstruction mentioned above.


## - TCM approach

According the Bob Flows and Philippe Sionneau acute renal failure can be divided into wind cold, wind heat, toxic poisoning, blood stasis, fluid exhaustion, lingering damp heat, and Qi and yin deficiency patterns.

In the presence of kidney deficiency wind cold may cause acute renal failure as the wind (yang) pathogen attacks the lung and the cold (yin) pathogen attacks the kidneys.

In cases associated with wind heat the treatment principle is to not only clear the lung and resolve toxins, but also to open water passages (a function belonging to the kidneys). This implies a dysfunction of the kidneys possibly associated with a deficiency of that organ as a lack of Qi would prevent the kidneys from performing their functions adequately.

Toxic poisoning is the equivalent a drug induced prerenal failure. In this case the accumulation of toxins will cause stagnation and impair the flow of Qi. The kidneys may thus become Qi deficient and lack the necessary energy to control the orifices.

For blood stasis cases the stagnation of Qi will not only cause the accumulation of toxins but the deficiency of the kidneys as well.

Fluid exhaustion cases are generally caused by some kind of dehydration or blood loss. In these cases the Qi and blood deficiency is obvious and thus follows the kidney deficiency as well.

With lingering damp-heat pattern, the accumulation of dampness may be explained by a spleen and kidney deficiency as is mentioned in the chronic renal failure section.

Qi and Yin deficiency may also cause acute renal failure but with profuse urination as is the case in lingering damp-heat patterns. The treatment principle is to strengthen spleen and kidney while tonifying Qi and yin. Again spleen and kidney deficiency are responsible for this pattern of acute renal failure.

Regardless of the pattern associated with this disease, kidney deficiency will either be present at the beginning or develop during the illness.

## B.   Effects of Cordyceps acute renal failure

The effects of Cordyceps sinensis on acute renal failure is very similar to that of chronic renal failure as the goal is to protect the kidneys from further damage and speed up the healing process of the injured parts.

Cordyceps seems to achieve this by increasing and redistributing epidermal growth factors of renal tissue, as well as speeding up the recovery of renal tubules, thus promoting the recovery of renal function. Other contributing functions include protecting the activity of the ATP enzyme, and reducing the oxidative degradation of lipids in order to protect cells from further damage.

## C.   Research studies on the subject

Guo Zhaoan has reviewed research published in China on the subject of acute renal failure. Tian Jin et al have induced renal tubular damage in rats and used Cordyceps sinensis to delay the occurrence of proteinurea and lower BUN levels. They showed that the fungus helps stabilize cellular lysosome in renal tubules by delaying its rupture. Similarly Zhen Feng et al showed that Cordyceps sinensis reduces the severity of acute renal tubular damage and promotes the recovery of renal damage ahead of time.

Liao et al also appeared to have shown the effectiveness of Cordyceps sinensis in reducing the flow of renal cortex mitochondrial calcium while protecting the activity of ATP enzyme thus improving renal function.

According to Georges M Halpern MD, PhD, a controlled study performed in China on people suffering from acute renal failure due to gentamicin showed that 89% of patients having taken Cordyceps sinensis made a complete clinical recovery after the sixth day while only 45% of those who took the western alternative to neutralize the toxicity of gentamycin achieved the same result.

These studies seem to indicate that acute renal failure can indeed benefit from the use of Cordyceps sinensis in its treatment.

Case in review: The case I am about to discuss covers both acute and chronic renal failure. Indeed this diabetic patient called me about two and a half years ago telling me that he had about three week to get his kidneys in working order or he would be placed on dialysis for the rest of his life. I put him on a high dose of Cordyceps sinensis and crossed my fingers. Three weeks latter his creatinine levels had gone down and his physician decided to wait another month before starting dialysis. One month later, creatinine levels were down again and dialysis was no longer an option. Unfortunately, about 4 months after that, the very same patient caught a gastrointestinal infection that send him to the emergency room. There, he suffered from an acute renal failure as a complication of the infection. He was placed on dialysis and thought he would probably remain there for life. After two months I suggested that he slowly get off dialysis as he was on Cordyceps again. His physician took him off completely and for one week his creatinine levels went back up. After one week, once the kidneys got the message that they had to work on their own again, creatinine levels went back down. As of today, the gentleman is doing well and has been dialysis free for over a year. As an added bonus he no longer takes any diabetes medication and has normal blood sugar levels.

# 7

# Cordyceps and Kidney transplant

## A.    Overview of kidney transplant

Kidney transplant plays an essential part in the west as it is the only way of surviving chronic renal failure when end stage renal disease is reached. Dialysis can postpone the inevitable, but only for some time and then kidney transplant is a must.

Antigenic differences between donor and recipient, the type of immune response developed by the receiver, and the immunosuppressive treatment administered to the patient are all factors affecting the success of the transplant and the length of survival of the sufferer.

According to Lawrence M. Tierney Jr. and coworkers, other factors affecting the success of the transplant are age of both donor and recipient, race of the receiving patient, length of time on dialysis, hyperlipidemia, hypertension, and cytomegalovirus infection.

Based on the number HLA (Human Leukocyte Antigen) mismatch the half life of the transplanted kidney may go from 6.8 years for 6 mismatches with a cadaver donor to 23.6 years for HLA identical sibling donor.

## B.   Effects of Cordyceps on kidney transplant

Cordyceps seems to be a first line of defense to kidney transplant. Indeed it seems to offset the side effects of drugs such as gentamicin and cyclosporine.

Gentamicin is an antibiotic drug that damages the kidneys. Cordyceps can provide appropriate protection against the type of damage rendered by this drug onto the kidneys.

Cyclosporine is an immunosuppressant drug used on organs transplant cases. Unfortunately, it damages the kidneys in the process by causing blood vessel constriction and cell damage at the glomerulotubular level which may in turn cause acute renal failure. Other side effects may include to development of diabetes mellitus, high blood pressure, cancer and a vulnerability to infections.

## C.   Research studies on the subject

In 1995 F. Xu and coauthors explored the amelioration of cyclosporine nephrotoxicity by Cordyceps sinensis in kidney transplanted patients. In this study, 69 kidney transplant patients were separated into two groups, group A (39 patients) receiving cyclosporine and placebo, and group B (30 patients) receiving Cyclosporin and Cordyceps sinensis. The results indicated that group B showed less signs of nephrotoxicity than group A and that the difference between the two groups increased with time. For instance BUN levels for group A were about 11mmol/l before treatment, 17mmol/l after 5 days of treatment and 20mmol/l after 15 days of treatment. Group B BUN levels were at about 10mmol/l before treatment, 14mmol/l at 5 days of treatment, and 15mmol/l after 15 days. Since the cyclosporine trough whole blood concentration did not change in both groups the researchers concluded that Cordyceps sinensis did indeed have a protective effect against cyclosporine nephrotoxicity.

Conclusion: This study seems to show that Cordyceps sinensis indeed has a positive effect on the outcome of kidney transplant in transplanted patients.

# Part C

# Cordyceps

# and

# Liver functions

- Overview of western liver functions

The liver, weighing about 1.5kg, sits just under the diaphragm and in the right hypochondrium and epigastrium. It is made up of two lobes, the left one being about six times smaller than the right. Lobules separated by fibrous strands and blood vessels make up the lobes.

The main functions of the liver include detoxifying a variety of substances, secreting bile, being involved in the metabolism of fats, proteins and carbohydrates, storing substances such as iron, copper, glucose and vitamins, producing plasma proteins and hosting blood cell production during the first trimester of fetal development.

Many toxic substances enter the blood stream via the intestines and travel to the liver to be transformed into harmless ones after undergoing a succession of chemical reactions. For instance alcohol and drugs such as Tylenol are modified so they become nontoxic in that organ.

The liver also converts ammonia to urea. It may also create toxic substances in its attempt to break down another.

Bile salts which are created from cholesterol in the liver are an essential component of bile which in turn plays an important role in the emulsification and absorption of fat from the intestines. Most are recycled in the liver. Bile also plays a role in the removal of the red blood cells breakdown byproducts. Indeed the heme portion of the hemoglobin that is released when a red blood cell is destroyed will be transformed to bilirubin and taken to the liver where it will be removed from the blood and excreted into the bile.

Gluconeogenesis, Glycogenolysis and glycogenesis are all part of the metabolism of carbohydrates and are performed by the liver. The breakdown of insulin and other hormones is also done by the same organ. In protein metabolism, the liver ensures the basics by converting substances such as lactic acid into amino acids like alanine thus providing the building block necessary for protein synthesis. The liver also synthesizes cholesterol and produces triglycerides.

As we have seen in a prior chapter, the liver also plays an important role in the immune system. Being made up of so many immunologically active cells allows the liver to filter out the antigens brought to it via the hepatic portal system.

## - Overview of oriental liver functions

According to Traditional Oriental medicine the functions of the liver include storing blood, allowing for the smooth flow of Qi, controlling the sinews, manifesting in the nails, opening into the eyes, and housing the hun.

As the liver stores blood during rest it also regulates the volume of blood going to the sinews and muscles during physical activity. When the blood moves to an area during activity, it nourishes the area and the person feels energetic. When the liver does not function adequately, the blood will not be delivered to the appropriate area and energy will be lacking. Proper blood flow will also ensure that skin and muscle are adequately nourished to fight off an external pathogenic attack as it plays a minor but none-the-less important factor in expelling the offending pathogen.

Menstruation cycles may also be affected by the liver ability to store blood. Problems will arise when blood is no longer moving as it should. Amenorrhea or oligomenorrhea may develop when the blood is deficient. When liver blood is excessive the menstrual flow will be heavy and the woman will bleed for a longer period of time.

The liver's ability the maintain the smooth flow of Qi throughout the body is essential for the proper functioning of the individual as it ensures that the organs will do what they are supposed to.

The proper flow of Qi also ensures the emotional well being of the individual. When the flow of Qi is impaired, the Qi stagnates and emotional symptoms such as frustration and depression develop. The reverse is also true. Frustration and anger will eventually lead to the stagnation of Qi and the impairment of liver function.

A proper Qi flow also helps the spleen and stomach with their digestive functions. When the Qi flows adequately the stomach can easily ripen and rot food while the spleen can extract the food-Qi. When Qi is stagnant, the stomach Qi will no longer be able to move downward and rebellious Qi with follow. Similarly the spleen will not be able to transform and transport and spleen Qi will not be able to flow upward. Diarrhea will follow. Bile flow may also be obstructed if liver Qi is stagnant.

Proper nourishment of the sinews by the blood that comes from the liver is essential for physical activity and the body's ability to move in general. If liver blood is deficient, not enough nourishment and moisture will reach the sinews and spasms, numbness, cramps or tremors might follow.

As the liver manifests in the nails, any deficiency of blood will prevent proper nourishment and moistening of the nails and thus will cause dry, brittle or concave nails.

The liver sends blood to the eyes for proper nourishment and moistening in order for the eye to see adequately. A deficiency of liver blood may lead to eye disturbances such as blurry vision, floaters, or color blindness.

The Hun or Ethereal Soul is housed in the liver. It is yang in nature and when death comes it will flow out of the body where it will survive as energy. Mental confusion and lack of direction in life may be associated with a weak liver blood and thus a weak Hun.

# 8

# Cordyceps and fatty liver

## A.  Disease mechanism

-  Western approach

Fatty liver disease is an accumulation of fat within hepatocytes that generally does not cause permanent damage. At this stage the liver becomes yellow and enlarged.

Fatty liver disease may be related to alcohol consumption, fat content in food consumed, body fat, hormones, or other unknown factors as the pathogenesis of this disease is not quite understood.

Nonalcoholic steatohepatitis (NASH) is a potentially more severe type of fatty liver disease associated with liver-damaging inflammation and, occasionally, the formation of fibrous tissue. This can eventually develop into either liver cancer, or cirrhosis which can in turn produce progressive, irreversible liver scarring.

- TCM approach

Traditional Oriental medicine appears to consider that fatty liver is caused by an accumulation of damp heat and phlegm in the liver. Very little information is available and no exact disease mechanism has been found to date.

## B.  Effects of Cordyceps on fatty liver

Cordyceps undeniably has protective and reparative effect on the liver. However the way in which this is accomplished remains unclear.

Some suggest that the fungus inhibits the synthesis of $TGF\beta_1 mRNA$ which in turn causes the slowing down of $TGF\beta_1$ and PDGF expression which in turn will reduce the deposition of collagen in the liver. As the disease develops collagen deposits increase as well. So reducing collagen deposits should slow down and possibly even stop the progression of this disease.

Furthermore, Cordyceps sinensis help normalize liver enzymes and thus improve overall liver function.

## C.  Research studies on the subject

In 2008 Wang Ting et al published a study on the effect of Cordyceps sinensis on alcoholic fatty liver in rats. 40 SD rats were used and divided into four equal groups. The normal group did not have fatty liver. The Model had fatty liver and received no treatment. The dietary control group had fatty liver and was placed on a restricted diet. The CS group had fatty liver had was given Cordyceps. The results for liver enzymes were as follows (standard deviation will be omitted for ease of reading): AST level for normal group was 180.3, Model group 240.3, Dietary group 220.3 and CS group 190.5. ALT levels were 59.3 for normal group, 90.6 for model group, 80.7 for Dietary group and 62.6 for CS group. Similar trends were observed for blood fat and HAI of liver cells.

Conclusion: This clearly indicates that Cordyceps sinensis is more effective than diet and even produces results close to those of normal subjects.

In 2006 Yang Zhao-xia and his colleagues studied the effect of Cordyceps sinensis on rat fatty liver and the possible molecular mechanism associated with it. 32 female rats were used and divided into 4 groups. Group B, made up of 8 rats, was the diet treated group and had a low fat diet. The remaining 24 rats were placed on a high fat diet. Group M (NAFL) was the model group and was sacrificed after eight weeks. The CS group was fed Cordyceps sinensis and the P (NASH) group was the pathology group given saline solution instead from the ninth week on. Liver histopathology, serum tumor necrosis factor alpha levels and UCP2mRNA expression were evaluated. B group showed no steatosis, inflammation necrosis or fibrosis while the M group showed steatosis (3.13), along with the CS group (2.88) and the P group (3.88). Inflammation necrosis only showed in the CS group (1.5) and the P group (2.88). Fibrosis was only present in the P group (1.13). Expression of UCP2 protein was 0.004 for group B, 0.026 for group M, 0.030 for group CS and 0.04 for group P, while that of the mRNA was 0.146 for group B, 0.416 for group M, 0.613 for group CS and 0.749 for group P. Serum TNFα was measured at 22.53 for group B, 28.20 for group M, 47.99 for group CS, and 67.99 for group P. Liver TNFα was measured at 0.006 for group B, 0.026 for group M, 0.039 for group CS, and 0.056 for group P.

Conclusion: The authors concluded that Cordyceps sinensis can inhibit steatohepatitis derived from nonalcoholic fatty liver disease by lowering serum and tissue TNFα, and reducing the over-expression of UCP2.

In 2003 Yu-Kan Liu and Wei Shen studied the inhibitive effect of Cordyceps sinensis on experimental hepatic fibrosis. For this experiment they used a normal control group, a model control group and a CS group. The model group was given $CCl_4$ and ethanol only while the CS group was also fed Cordyceps sinensis. Results indicated that serum ALT, AST, HA and LN were significantly dropped in the CS group when compared to the model group (86.0 vs. 224.3, 146.7 vs. 272.6, 202.0 vs. 316.5 and 50.4 vs. 59.7 respectively). Other criteria were used and showed similar results.

Conclusion: The authors concluded that Cordyceps sinensis could indeed inhibit hepatic fibrogenesis derived from chronic liver injury as well as retard the development of cirrhosis.

From these studies we can see the clear advantage of using Cordyceps sinensis to reduce the speed of liver degeneration and maybe even to reverse possible damage.

# 9

# Cordyceps and Cirrhosis

## A.   Disease mechanism

-      Western approach

Cirrhosis represents the end-stage of alcoholic liver disease. It is associated with recurring bout of drinking or hepatitis where the liver suffers irreversible damage. In the beginning stage of the disease the liver shows signs of having fine, regular nodules on its surface. As the illness progresses the nodules become larger and more uneven in size and shape. Over time these nodules may constrict the hepatic vein thus reducing blood flow out of the liver and causing portal hypertension, extra hepatic portosystemic shunts (bypass of the liver by the circulatory system) or the bile flow from the liver to the duodenum may become obstructed. Eventually splenomegaly and ascites will develop. Ultimately liver complications develop and the patient may suffer from liver cancer or chronic active hepatitis.

There are three main types of cirrhosis, namely post-necrotic cirrhosis, primary biliary cirrhosis and alcoholic cirrhosis.

In post-necrotic cirrhosis, which represents 10 to 30% of cirrhosis cases, dead cells are replaced with fibrous nodules of different sizes. The liver thus acquires an irregular shape. It can be due to Hep B or C, an autoimmune disorder or a response to a toxin.

According to Carol Mattson Porth, primary biliary cirrhosis refers to the inflammation and scarring of small bile ducts within the liver, inflammation of some portal fields, and liver scarring. It represents 2 to 5% of cirrhosis cases and may be due to an autoimmune response.

In alcoholic cirrhosis the metabolism of the alcohol creates chemicals capable of attacking some liver membranes. Areas of cell damage appear to be concentrated around the central vein and thus where the alcohol metabolism occurs. This area has been determined to contain less oxygen. As such it has been speculated that this lack of oxygen may be at least partially responsible for the observed cellular damage. If alcohol consumption is stopped, damage will continue for months, but fat accumulation will disappear within weeks and inflammation will eventually resolve after that. Only the presence of scarring and fibrous tissue will be detected and new cells ultimately develop.

## - TCM approach

Traditional Oriental medicine blames a fatty liver on the accumulation of Damp and heat in the liver and gallbladder.

When damp-heat is due to internal causes the condition is chronic and generally comes from overeating fatty foods and drinking alcohol for a long time. When damp-heat is associated with external causes the condition is acute and generally comes from a hot humid environment or contaminated food.

When dampness is present the flow of Qi becomes sluggish and heat can develop. This pattern may worsen with the consumption of rich, fatty foods and alcohol.

In "The Treatment of Modern Western Medical Diseases with Chinese Medicine" the authors also describe cirrhosis "zheng jia" or mass formation and accumulation, jaundice, emaciation, and edema.

## B.    Effects of Cordyceps on cirrhosis

The effects of Cordyceps sinensis on cirrhosis have been demonstrated. It seems to protect the liver against further injury by stopping hepatocyte degeneration, but also decreases the amount of collagen deposit and fibrous septa present on the liver. It also appears to decrease the size and number of varicosities.

According to Georges M Halpern MD, PhD, Japanese researchers have also concluded from their study that the fungus increases the metabolism of the liver thus increasing the ATP levels which in turn may promote cell repair.

Cordyceps also appears to reduce the pressure in the hepatic portal vein which generally accompanies Cirrhosis.

Unfortunately the exact mechanism by which this happens is still unclear as it remains poorly researched. We may speculate that similar mechanisms than the ones suggested for fatty liver disease may be at play here as well.

## C.    Research studies on the subject

In 1996, Wang Yaojun et al studied the therapeutic effects of Cordyceps sinensis on decompensated cirrhosis. For one month, 18 patients in the treatment group were given Cordyceps sinensis while 16 patients in the control group were given Gantaile. In the treatment group both ALT and bilirubin serum levels decreased while albumin levels increased. Results were as follows: treatment group ALT before 82.5, after 47.3; control group ALT before 80.7, after 65.9; treatment group bilirubin before, 55.3, after 18.9; control group bilirubin before 50.8, after 46; treatment group albumin before 26.6, after 30.9; control group albumin before 26.9, after 27.6. Ascites also disappeared in a shorter amount of time and symptoms improved faster in the treatment group. The authors mentioned that no adverse side effects were observed

Conclusion: The authors concluded that Cordyceps sinensis can very possibly cure cirrhosis.

In 1994, Xu Lieming et al published a study on the effect of treating post hepatic cirrhosis with semen Persicae extract and Cordyceps sinensis. Patients were injected with 1.5g of semen Persicae extract every other day and given 4.5g of Cordyceps orally on a daily basis for a period of 3 months. Upon examination, the hardness and color of the liver had improved, while varicosity was decreased along with edema and ascites. Hepatocyte degeneration was halted and fibrous septa reduced in most cases. Collagen deposits were decreased in more than half the patients.

Conclusion: The authors concluded that both herbs seemed to be appropriate treatment for cirrhosis as they appeared to inhibit fibrous tissue hyperplasia in the liver.

Overall research seems to indicate that Cordyceps sinensis is a valuable ally in the fight against cirrhosis.

Case in review: One of my patients has been treated with Cordyceps sinensis and other Chinese herbs for his cirrhosis for over a year. He had no prior history of drinking and his disease was of unknown origin. He has been stable now for over a year and his liver is functioning well.

# 10

## Cordyceps and Hepatitis

## A.    Disease mechanism

-      Western approach

Hepatitis is the inflammation and death of liver cells. Dead cells generally occur in patches and may affect an entire lobe. It is however reversible.

Hepatitis can be the intermediate stage between fatty liver and cirrhosis. However there are different types of hepatitis.

Hepatitis A is caused by a small, envelop lacking, RNA containing virus and is generally benign and self-limited. It is generally caught from infected shellfish, milk or water. Very rarely does Hepatitis A evolve into fulminant hepatitis (liver cell necrosis leading to liver failure) and death from liver failure.

Hepatitis B is associated with a double stranded DNA virus. The disease can evolve in acute hepatitis or chronic hepatitis which can in turn become cirrhosis, or fulminant hepatitis. The patient may also become a carrier. Hep B is transmitted through blood and sexual contact as most body secretions contain the virus as well.

Hepatitis C is associated with a single stranded RNA virus and is transmitted through blood. As it has also been detected in bodily fluids, it can also be transmitted via high risk sexual contacts. Most cases are initially asymptomatic but the disease eventually develops into acute hepatitis, chronic hepatitis or the patient becomes a carrier.

Hepatitis D is caused by a defective RNA virus and is transmitted through blood and sex. The disease can develop into acute or chronic hepatitis but require the help of the Hep B virus. Once it has infected a person already contaminated with the Hep B virus it will generally increase the severity of the Hep B infection, sometimes to the point of converting it to a fulminating one.

Hepatitis E is caused by a single stranded RNA virus lacking an envelope. Transmitted by the fecal-oral route, even though it does not produce chronic hepatitis or cause the patient to become a carrier, Hep E tends to develop into fulminant hepatitis and thus cause death.

When acute hepatitis persists for longer than 6 months the patient is said to have developed chronic hepatitis. This disease can either be persistent or active. When persistent, hepatitis is generally benign as little cellular death is detected. It may however develop into the active form which does exhibit progressive cellular necrosis.


-    TCM approach

Traditional Oriental medicine views hepatitis as being caused by an external condition, overeating, excessive drinking, internal damage due to the seven emotions and aging.

In the case of hepatitis the evil Qi goes deep in the blood level and it is damp and hot in nature. Not only does it damage the blood but it also affects the Qi and thus causes Qi and Blood stagnation. Any other input of damp-heat into the body will make the situation worse.

Internal causes can be narrowed down to overeating of hot, spicy, sweet or fatty foods, drinking alcohol and emotional stress. The kind of diet described above will damage the spleen and thus cause dampness which will in turn cause heat formation. When due to emotional stress the liver will be affected. The stagnation of Qi in this organ will cause the formation of heat and will cause the liver to overact on the spleen. The resulting spleen deficiency will then create dampness which will join with the heat to make damp-heat.

Aging is generally associated with spleen deficiency. The righteous Qi thus becomes weak as well and can fight off the evil Qi effectively and invasion can easily occur.

Damp-heat will cause Qi and blood stasis as well as phlegm, but heat will cause yin, blood, and fluid damage and will therefore lead to yin and/or blood deficiency. Dampness is however the dominating factor and the disease may as such remain minor for years. Only when the righteous Qi becomes too weak does the disease become severe.

## B.    Effects of Cordyceps on hepatitis

The effects of Cordyceps sinensis on Hepatitis are many. It helps regulate cellular immunity and thus search and destroy infected cells. It also decomposes collagen and softens the liver. It promotes lever cell regeneration and helps reduce the fatigue associated with hepatitis cases. It lowers ALT levels and increases albumin. It seems to improve overall liver function and relieves symptoms associated with the disease.

Again the exact mechanisms by which these improvements occur are not thoroughly understood.

## C.    Research studies on the subject

In 2006 Liang Huijing and his colleagues published a study on the effect of Cordyceps sinensis and Anti HB positive placenta for the treatment of hepatitis B. Two groups were established. The control group (CG) had 28 cases while the treatment group (TG) had 30. Both groups were fed three Ganli capsules and 0.3g of vitamin C 3 times daily, while the treatment group also received half a bottle of Cordyceps sinensis liquid and 30g of Hep B surface antibody positive placenta per day for a period of 3 months. The results showed a marked improvement in the treatment group and were as follows: fatigue 89.3% (TG), 53.3% (CG); anorexia 91.75% (TG), 57.7% (CG); disappearance of abdominal distention 94.4% (TG), 68.4% (CG); liver lump retraction 31.6% (TG), 20.0% (CG); spleen lump retraction 35.3% (TG), 16.7% (CG); ALT normalization rate 82.1% (TG), 40.0% (CG); negative conversion rate of HBeAg 42.3% (TG), 7.1% (CG); and negative conversion rate of HBV-DNA 44.4% (TG), 6.9% (CG).

Conclusion: The researchers concluded that these results clearly demonstrated an improvement in the treatment of hepatitis B when Cordyceps sinensis and Hep B surface antibody positive placenta were used.

In "Healing Mushrooms", Dr Halpern describes yet another study where 83 patients suffering from hepatitis B and ranging in age from 2 to 15 years, were treated with Cordyceps sinensis for 3 months. The results indicated that 33 of those patients showed a complete conversion of antibodies to the disease while the number of antibodies positive to the virus had reduced in 47% of the cases. He further suggest that as impressive as these results were, they could have been even better if the subjects had been older and with a mature immune system.

In 2000, Huan-Yu Gong and his colleagues published a study on the effects of Cordyceps sinensis on T-lymphocyte subsets and hepatic fibrosis in patients with chronic hepatitis B. The trial group (25 patients) was to take 1g of Cordyceps sinensis 3 times a day along with Vitamin E, C and K, diammonium glycerrihizinate and panangin. The control group (18 patients) took all of the above but omitted the Cordyceps sinensis. The patients were tested over a 3 months period. For the trial group the $CD_4$ percentage was 37.47 before treatment and 39.64 after treatment while the $CD_8$

percentage was 28.8 before treatment and 27.3 after treatment. For the control group the $CD_4$ percentage was 37.87 before treatment and 38.29 after treatment while the $CD_8$ percentage was 28.3 before treatment and 27.27 after treatment. The hyaluronic acid levels in $\mu g.L^{-1}$ were 507.68 before treatment and 264.92 after treatment for the trial group, and 498.35 before treatment and 455.33 after treatment for the control group. It is worth noting that elevated hyluronic acid levels are consistent with liver diseases such as hepatic fibrosis and cirrhosis. Similarly patients suffering from these diseases also exhibit high levels of procollagen III. In this study the levels of PCIII in $\mu g.L^{-1}$ were 196.65 before treatment and 121.58 after treatment for the trial group, and 221.30 before treatment and 187.347 for the control group.

Conclusion: The investigators concluded that Cordyceps sinensis indeed helped reduce hepatic fibrosis, as demonstrated by the decrease in both PCIII and hyluronic acid in the trial group. Furthermore Cordyceps sinensis seemed to adjust cellular immune functions as it did increase $CD_4$ levels and decrease $CD_8$.

In 1990, Liangmei Zhou and his colleagues published an article on the short-term curative effect of cultured Cordyceps sinensis for chronic hepatitis B. The study involved 33 patients who were given 3.75g of Cordyceps sinensis daily for a period of 3 months. Among others thymol turbidity tests (TTT), SGPT, and thymol flocculation tests (TFT) were run. 32 out of the 33 patients started with abnormal TTT. At the end of the three months 31.25% of the patients had returned to normal levels, 40.6% showed obvious improvement, and 28.15% showed no improvement. TFT levels were abnormal in 32 cases at the beginning of the trial and had returned to normal in 31.25% of the cases, showed obvious improvement in 31.25% of the cases, and no improvement in 37.5% of the patients after 3 months of treatment with Cordyceps sinensis. SGPT levels were abnormal in 14 cases at the beginning of the trial, 64.28% returned to normal, 14.28% showed obvious improvement and 21.42% showed no improvement at the end of the trial.

Conclusion: The authors concluded that the fungus did improve liver function and that it should therefore be used as a medicine for treating chronic hepatitis B.

These studies appear to indicate that Cordyceps sinensis, alone or in the presence of other compounds, indeed improves the chances of recovery of patients suffering from hepatitis.

# Part D

# Cordyceps

# And

# Lung functions

- Overview of western lung functions

The lungs, shaped in the form of a cone, can be found between the diaphragm and a point slightly above the clavicle. At the same time they are encased within the rib cage. Primary bronchi and pulmonary blood vessels enter the lungs on the medial aspect. The left lung has two lobes while the right one has three, and within each lobe enter secondary bronchi. Lobes are also made up of bronchopulmonary segments that receive their air from tertiary bronchi. Each of these segments is made up of countless tubes of decreasing diameter that together they make up the bronchial tree. Each tube ends with an alveolus where the gas exchange occurs.

In the western model the lungs are responsible for air distribution and gas exchange. During air distribution the air travels through the bronchial tree to the alveoli. Once the air reaches the alveolus, gas exchange occurs between alveolus and the surrounding blood capillary network. The area where the alveoli and capillaries come into contact in called the respiratory membrane. This membrane is responsible for the very rapid circulation of gases between air and blood. Oxygen moves into the blood stream while carbon dioxide exits.

## -    Overview of oriental lung functions

As described in "The Foundations of Chinese Medicine" Traditional Oriental medicine considers the functions of the lungs to include governing Qi and respiration, controlling channels and blood vessels, controlling dispersing and descending, regulating water passages, controlling skin and hair, opening into the nose, and housing the Corporeal Soul.

The lungs extract clean Qi for the body to use after it combines with the food Qi harvested by the spleen. The lungs are said to govern respiration because they take in pure Qi and breathe out dirty Qi. If this function is not fulfilled appropriately the body will not get the nourishment it needs for proper functioning. The lungs also govern Qi in the sense that the air taken in from the environment combines with the food Qi from the spleen to make Zong Qi in the lungs. From there the Zong Qi is sent all over the body to provide appropriate nourishment.

The Zong Qi obtained from the lungs is vital to the proper functioning of the heart and thus blood circulation. So even though the heart controls blood vessels, the lungs are essential at keeping them healthy. As both Qi and blood travel in blood vessels and channels, and since the lungs govern the Qi they will thus control the circulation of Qi in blood vessels and channels.

The lung ensures that Wei Qi is dispersed throughout the whole body along with body fluids that it delivers as a fine mist. The Wei Qi is sent between skin and muscle keeping the skin warm and resisting external pathogenic attack. Body fluids are also sent to the skin, and as mist they moisten the skin and regulate the opening of pores and the subsequent sweating.

The lungs, being the highest organs in the body, have to send Qi down as there is nowhere else to go. As such they control descending. Lung Qi has to descend to communicate with the kidney Qi, and body fluids need to descend as well to reach both kidney and bladder. Lung Qi must also descend to reach the large intestine in order to provide the energy necessary for a bowel movement.

Because the lungs are responsible for distributing water in the form of mist all over the body as well as sending it down to the kidney and bladder for processing and controlling the excretion of body fluids through sweat and urine, the lungs can be considered to regulate water passages.

As lungs receive refined fluids from the spleen and send them to the skin throughout the body, both skin and hair are nourished and moistened. Because the lungs are responsible for the skin and hair's proper nourishment and moistening they control skin and hair.

The nose is where air enters the body to get to the lungs. As such the lungs are said to opens into the nose. Furthermore, the state of the lung Qi will determine whether the nose remains open and able to send air down or not. For instance, a weak lung Qi may cause nasal congestion or sneezing or even loss of smell thus demonstrating the relationship between lung and nose.

The lung houses the Corporeal Soul or the physical part of the Soul. It is closely related to Jing which is in turn essential for good health. The Corporeal Soul is responsible for movement and is also related to sadness and grief. As such treatment of the lung is of the utmost important when dealing with these types of emotions.

# 11

## Cordyceps and bronchitis

## A.  Disease mechanism

-  Western approach

In this discussion, bronchitis refers to chronic bronchitis a disorder that is caused by airway inflammation. Edema and excessive cellular development of submucosal glands as well as excessive mucus secretion in the bronchial tree are present. The patient must have suffered from a chronic productive cough for at least 3 months and for at least two consecutive years without the presence of other diseases before he/she can be diagnosed as having chronic bronchitis.

Air normally moves through the trachea and bronchi to the alveoli. Cartilage supports the upper part of the air pathway and is then replaced with smooth muscle in the lower airways. These smooth muscles are innervated by autonomic nervous system and thus control air flow. When cholinergic receptors are stimulated (parasympathetic), the vagus nerve sends an impulse to the smooth muscles that will cause an increase in bronchial constriction. When B2-adrenergic receptors are stimulated (sympathetic), bronchodilation is induced.

However bronchial smooth muscles are also affected by histamines and other inflammatory triggers which can cause bronchial constriction and inflammation.

- TCM approach

Chronic bronchitis is one of the Chronic Obstructive Pulmonary Diseases (COPD). As such traditional Oriental Medicine attributes similar mechanisms for chronic bronchitis and other COPD.

Anshen Shi describes chronic bronchitis as having both extrinsic and intrinsic causes. Exterior pathogens invade the body through the skin, mouth and/or nose and attack the lungs where they upset their dispersing and descending functions. Viruses, bacteria, excessive consumption of tobacco, alcohol, or spicy food may trigger such as response. The airways may become obstructed and the lung Qi may rebel, causing coughing, breathlessness and possibly wheezing. Intrinsic causes may be a spleen dysfunction leading to phlegm accumulation and thus airway obstruction, kidney failing to receive Qi causing stagnation in the lungs and thus rebellious Qi moving upward, kidney Yin deficiency with empty fire causing lung fluid injury and thus lung Qi failing to descend.

The interaction between intrinsic and extrinsic as well as extrinsic factors together creates a vicious circle that promotes exacerbation during remissions. Exacerbation episodes are either caused by exterior wind-cold activating the dormant congested fluids or by exterior wind-cold with phlegm-heat accumulation. But a deficient root is always present. During remission phlegm-damp accumulation in the lungs, phlegm-heat obstructing the lungs, lung deficiency, kidney deficiency, and/or spleen and kidney Yang deficiency may be present.

## B.   Effects of Cordyceps on bronchitis

It has been suggested that Cordyceps sinensis regulates the immune system and that as such it decreases the inflammation present on the lungs thus allowing for better air flow.

Indeed this fungus has been known to Chinese medicine for hundreds of years for its curative properties on the lungs. It is used to stop coughing, strengthening the lungs and resolving phlegm.

## C.   Research studies on the subject

In this section we will be discussing Studies involving Cordyceps sinensis and Cordyceps militaris as it is considered by some authors as the same fungus with similar functions. Only the place where they originally grew appears to differentiate them.

In 2004, Guozhong Gai et al published an article on the efficacy of Cordyceps militaris capsules in the treatment of chronic bronchitis in comparison to Jinshuibao capsules. 315 patient received 3g of Cordyceps militaris per day for 2 months while 110 patients received 3g of Jinshuibao par day for 2 months. All patients were diagnosed with chronic bronchitis, lung weakness and kidney yang deficiency. They observed cough, sputum, lung and kidney tonifying effect, asthma, and inflammation. In the Cordyceps group the treatment was excellent in 24.1% of the cases, effective in 28.6% and had no effect in 7.3% of the cases. When given Jinshuibao capsule, 15.4% of patients had excellent results, 31.3% showed the treatment to be effective, and 29.1% showed no improvement. Before treatment 156 patients reported very bad cough, 143, bad cough, 15, some cough, and 1 no cough. After the 2 months of Cordyceps treatment only 6 patients reported very bad cough, 14, bad cough, 125, some cough, and 170, no cough. Before the treatment 75 patients reported having a lot of sputum, 160 having sputum, 77 having some sputum, and 3 having no sputum. After the 2 months Cordyceps treatment only 1 patient reported having a lot of sputum, 15 having sputum, 123 having some sputum, and 176 having no sputum. Dyspnea and wheezing rales showed similar results.

Conclusion: The authors concluded that Cordyceps militaris is safe and effective in treating chronic bronchitis.

In 1988, Yunchun Fang and his colleagues studied the effects of Cordyceps sinensis on chronic bronchitis. The treatment group was made up of 65 patients who received Cordyceps sinensis granules 2 to 3 times a day for 2 to 3 months to be repeated 4 times. The control group was made up of 42 patients that were given negundo chastetree oil drop pill or other traditional Chinese formula. For cough, the overall effectiveness was 93.85% for the treatment group and 60.06% for the control group while the overall ineffectiveness was 6.15% for the treatment group and 39.39% for the control group. For phlegm the overall effectiveness was 85.96% for the treatment group and 54.84% for the control group while the overall ineffectiveness was 14.03% for the treatment group and 45.16% for the control group. For acute attacks the overall effectiveness was the same for both groups. However, the efficacy of the treatment in chronic protracted cases appeared to be better for the treatment group with an overall effectiveness of 97.5%. The control group only had an overall effectiveness of 50%. For non emphysema patients the overall effectiveness for the treatment group was 96% while it was 55% for the control group. For emphysema patients the overall effectiveness for the treatment group was 70.97% while it was 22.22% for the control group. For heart disease patients no significant improvement was seen in both groups. It is worth noting that 2 patients had to drop out of the study because of frequent diarrhea due to the treatment.

Conclusion: The authors concluded that Cordyceps sinensis was helpful in the cases of chronic bronchitis except for those who exhibited concurrent heart disease. They also suggested that a treatment of Cordyceps sinensis for patients suffering from spleen, kidney and heart Yang deficiencies was not suitable. However the data they offer does not support nor negate this statement.

The research in the area of Cordyceps as a possible treatment for chronic bronchitis seems to indicate that indeed this fungus is effective and should be considered seriously when treating this disease. However it is also important to be aware of its possible limitations when it comes to using it in patients suffering from chronic bronchitis with coexisting heart disease, or the development of side effects such as diarrhea.

# 12

## Cordyceps and COPD

## A.   <u>Disease mechanism</u>

-      Western approach

COPD or chronic obstructive pulmonary disease is described as the chronic or recurrent airflow obstruction within the lungs. Its most common cause is smoking but other risk factors such as exposure to airborne toxins in the workplace or congenital deficiency of $\alpha_1$-antitrypsin may be responsible for the development of this disease.

The mechanisms involved in COPD consist of inflammation and the excessive mucus secretion within the pulmonary airways which cause airflow obstruction as well as an imbalance between alveolar ventilation and pulmonary capillary blood flow, the loss of alveolar tissue which in turn reduces the surface area available for gas exchange, the fibrosis of the bronchial wall as well as the loss of elastic lung fibers which reduce the expiratory flow rate and may cause airway collapse, edema, and the excessive cellular development of the submucosal glands.

-      TCM approach

As discussed in the previous chapter, COPD includes chronic bronchitis .As such, in traditional Oriental medicine, similar mechanisms are going to be at play when COPD is going to be talked about.

Anshen Shi describes COPD as being the consequence of a variety of lung dysfunctions or diseases with a insidious onset and slow progression. As such it can be considered as a deficiency responsible for the increased susceptibility to exterior pathogens.

COPD is thought to have both extrinsic and intrinsic causes. Skin, mouth and/or nose are the port of entry for exterior pathogens as they go attack the lungs and upset their dispersing and descending functions. Viruses, bacteria, excessive consumption of tobacco, alcohol, or spicy food may trigger such a response. The airways may become obstructed and the lung Qi may rise up, causing coughing, breathlessness and possibly wheezing. Intrinsic causes may be a spleen dysfunction leading to phlegm accumulation and thus airway obstruction, kidney failing to receive Qi causing stagnation in the lungs and thus rebellious Qi moving upward, kidney Yin deficiency with empty fire causing lung fluid injury and thus lung Qi failing to descend.

As with chronic bronchitis, the interaction between intrinsic and extrinsic as well as extrinsic factors together creates a vicious cycle that promotes exacerbation during remissions. Exacerbation episodes are either caused by exterior wind-cold activating the dormant congested fluids or by exterior wind-cold with phlegm-heat accumulation. But a deficient root is always present. During remission phlegm-damp accumulation in the lungs, phlegm-heat obstructing the lungs, lung deficiency, kidney deficiency, and/or spleen and kidney Yang deficiency may be present.

## B.    Effects of Cordyceps on COPD

Cordyceps sinensis seems to affect the lungs and thus the outcome of COPD in different ways. It seems to inhibit tracheal contraction, has an anti-inflammatory effect, and appears to increase the gas transport in the alveoli.

As mentioned in the previous chapter, this fungus has been known to Chinese medicine for hundreds of years for its curative properties on the lungs. It is used to stop coughing, strengthening the lungs and resolving phlegm. It is also known to replenish the kidneys, and enhance body immunity.

## C.    Research studies on the subject

In 2007, Cai-hong Guan and his colleagues published a study on the effects of Cordyceps sinensis on airway inflammation and pulmonary functions in Rats with COPD. The research is done on 3 groups. Both the model group and the treatment group smoke while the contrast group does not. The treatment group also receives 6ml of Cordyceps sinensis solution per day. The total number of leukocytes ($x10^6$/L) was 113.57 for the contrast group, 239.29 for the model group and 137.14 for the treatment group. The total number of macrophages ($x10^6$/L) was 2.33 for the contrast group, 36.90 for the model group and 7.90 for the treatment group. The total number of neutrophilic granulocytes ($x10^6$/L) was 100.86 for the contrast group, 160.70 for the model group and 102.73 for the treatment group. The number of times coughed in 10 minutes was 6.57 for the contrast group, 27.14 for the model group and 14.00 for the treatment group. The average number of alveolus was 60 for the contrast group, 37 for the model group and 50 for the treatment group. The pulmonary compliance (ml/cm $H_2O$) was 0.315 for the contrast group, 0.182 for the model group and 0.265 for the treatment group.

Conclusion: The authors concluded that Cordyceps sinensis improves the immune response and pulmonary function of rats suffering from COPD.

In 2004, Haoyu Qian published a study on the therapeutic effects of Cordyceps sinensis on COPD. 40 patients were part of the treatment group and 20 healthy adults made up the control group. Patients were given 1g of Cordyceps sinensis 3 times daily for a period of one month. Forced vital capacity (in L) was determined. For the control group is was 3.58, while the treatment group had an FVC of 2.39 before treatment and 2.89 after treatment. The forced exhalation volume (in L) was also determined. For the control group is was 2.43, while the treatment group had an FEV of 1.45 before treatment and 1.99 after treatment. $PaO_2$ (in mmHg) was evaluated as well. For the control group is was 99.8, while the treatment group had a $PaO_2$ of 86.5 before treatment and 94.3 after treatment. No remarkable changes were observed for the $PaCO_2$, the $SaO_2$ and the pH between the control and before and after treatment measurement of the treatment group. Endothelin (ET-1) levels (in ng/L) were at 32 for the control group, 71 before treatment and 57.9 after, for the treatment group.

Conclusion: The author concluded that because improvement was clear for FVC, FEV, and $PaO_2$, that Cordyceps sinensis was indeed helpful in treating COPD. They further concluded that because ENT-1 levels were also affected by Cordyceps sinensis, the fungus must decrease hypertension in the pulmonary small vessels, and thus help regulate hypertension in the pulmonary artery and reduce the risks of developing chronic pulmonary heart disease and right heart failure. However they agreed that further studies would be required to confirm their conclusion.

These studies may indeed indicate that Cordyceps sinensis helps with the treatment of COPD. Even if its mechanism of action is not completely understood, its impact on respiration is undeniable in both animal and human studies.

Note: a friend of mine with COPD decided to take Cordyceps sinensis to see if it could help him. He claimed to feel a marked improvement in his ability to breathe within 48 hours. No actual measurement of lung capacity were obtained though so this is a purely subjective but none-the-less interesting story.

# 13

## Cordyceps and asthma

### A.    Disease mechanism

-    Western approach

Asthma, or bronchial asthma, is a chronic inflammatory disease affecting the lung's air passages. More specifically, it is defined as an airway obstruction with inflammation, that is reversible either spontaneously or with treatment, along with increased airway sensitivity to possible allergens.

Two types of triggers are responsible for asthma attacks. Bronchospastic triggers do not affect airway sensitivity. Only when the responsiveness is already present does the asthma sufferer become affected by cold air, exercise, emotional distress, or contact with irritants such as fumes and cigarette smoke. Inflammatory triggers create an inflammatory response.

There are two types of responses to the different triggers discussed above. When there is an early response, symptoms of an asthma attack develop within 10 to 20 minutes of exposure to an allergens and disappear spontaneously after 60 to 90 minutes. In this case mast cells present on the mucosal surface of the airways release chemical mediators after the allergens bind to them, leading the bronchospasms. When symptoms develop 3 to 5 hours after exposure to the trigger and potentially last from a few days to a few weeks, it is said to be a late response. Here inflammation and airway sensitivity are involved and extend the attack thus exacerbating the condition further. Mast cells, macrophages and epithelial cells release inflammatory mediators that stimulate the movement and activation of other inflammatory cells which in turn produce epithelial injury, edema, changes in mucociliary function, reduced phlegm expulsion from the airways, and increased airway sensitivity.

During an asthma attack bronchospasms effectively cause the narrowing of the airways. Edema of the bronchial walls and mucus plugging also occur. Air output becomes labored and air may become trapped behind the mucus blockage thus causing lung hyperinflation. The functional residual capacity is increased, and inspiratory reserve capacity and forced vital capacity are decreased, causing the asthma sufferer to breathe close to his total lung capacity. The use of accessory muscles and extra energy becomes necessary to sustain proper ventilation and gas exchange. Dyspnea and fatigue ensue. Cough is less effective. Alveolar ventilation decreases. The imbalance between alveolar ventilation and pulmonary capillary blood flow then leads to decreased oxygen level and increased carbon dioxide in the blood. From then on, pulmonary vascular resistance may occur and cause pulmonary hypertension and thus excess strain on the right heart. Systolic blood pressure may fall during inspiration.

## -    TCM approach

In traditional Oriental medicine, asthma is caused by exterior pathogens activating the dormant phlegm in the lungs. It can also be caused by internal damage due to the seven emotions, poor diet, immaturity, aging, and drug therapy.

Phlegm may be a consequence of the lungs' failure to distribute and disperse fluids, or of the spleen's inability to transform and transport. The kidneys may also be responsible if they fail in their function of steaming and transforming fluids. When any of the above mentioned explanations occurs phlegm accumulates and stays dormant in the lungs until weather changes or other triggers stir it up.

During an attack phlegm obstructs the lungs and may be accompanied by either hot or cold patterns.

In the remission phase the condition goes from an excess to a deficiency. Lung Qi deficiency leads to the inability to distribute fluids and consolidate the surface thus leading to the formation of phlegm and the increased vulnerability to exterior pathogens.

A deficient spleen may also generate phlegm and increase the amount present in the lungs.

Kidney deficiency may also affect phlegm. When the kidney is yang deficient, it fails to process fluids properly. When it is Yin deficient with an empty fire it consumes body fluids. In both cases phlegm is produced.

# B.    Effects of Cordyceps on asthma

In terms of western medicine Cordyceps sinensis seems to help reduce the inflammation present in the airways of asthma patients. It may also help with the restoration of the airways. It may also increase the production of adrenal cortical hormones, and help expend bronchial smooth muscles.

In terms of Traditional Oriental medicine Cordyceps sinensis strengthens the kidneys and lungs, and tonifies the spleen, thus preventing the formation and accumulation of phlegm in the lungs.

# C.    Research studies on the subject


In 2007, Ningqun Wang and his colleagues published a study on the effect of Cordyceps sinensis capsules on the airway inflammation of asthmatic patients. A control group of 30 patients was given inhaled corticosteroids and β-agonist when needed. Depending on the level of treatment required the patients were to take bronchial dilator as needed. If that stopped working they were to be placed on low dose glucocorticoids plus bronchodilator when needed, then moderate doses of glucocorticoids plus bronchodilator when needed, then high dose glucocorticoids plus bronchodilator when need, and finally high dose of glucocorticoids plus bronchodilator regularly. A treatment group of 30 patients was given the same regimen but was also prescribed 5 capsules of Cordyceps sinensis to be taken 3 times daily after meals for 2 months. The results showed an increase in IgG (IgG/mg.mL$_{-1}$) from 42.49 before treatment, to 52.28 after treatment for the treatment group, and 49.85 before treatment and 66.94 after treatment for the control group. IgE (IgE/ng. mL$_{-1}$) was 37.56 before treatment and 27.01 after treatment for the treatment group, and 27.91 before treatment and 36 84 after treatment for the control group. MMP-9 (MMP-9/ ng. mL$_{-1}$) was 603.60 before treatment and 417.17 after treatment for the treatment group while is was 473.99 before treatment and 517.78 after treatment for the control group. Other makers such as sICAM, L-4 and IFN-γ were also used and the results were consistent with the ones described above.

Conclusion: The authors concluded that Cordyceps sinensis may adjust the balance of Th1/Th2 (T helper) cells, reduce L-4, decrease the adhesion molecule and IgE production, leading to reduced airway inflammation. Cordyceps sinensis may also improve the restoration of the airways.


In 2006, Juan Du and his colleagues published a study on the effects of Cordyceps sinensis on the cellular immune functions of asthmatic children. Bothe treatment and control groups (40 patients each) were given inhaled glucocorticoids but the treatment group was also given an infusion of Cordyceps sinensis for a period of 3 months. At the end of the 3 months 52.5% of the control group and 75% of the treatment group had no symptoms, normal physical signs and lung function, 27.5% of the control group and 20% of the treatment group still experienced occasional attacks but showed marked improvement, while 20% of the control group and 5% of the treatment group showed no improvement or aggravation of the attacks.

Conclusion: The authors concluded that the use of Cordyceps sinensis in the treatment of child asthma was advisable as it increases the efficacy of the glucocorticoid treatment.

These studies seem to indicate that Cordyceps sinensis does indeed help with the treatment of asthma in both adults and children.

# Part E

# Cordyceps

# And

# cardio-vascular functions

- Overview of western cardio-vascular functions

There are three main functions associated with the cardiovascular system. It delivers oxygen and nutrients as well as removes waste to and from the different organs and tissues of the body. It also plays a role in hormone transport and heat distribution throughout the body.

There are three parts to the cardio-vascular system: the heart which is located in the chest in front of the spine ,behind the sternum and between the lungs; the systemic circulation and the pulmonary circulation.

A combination of two pumps, the heart is made up of four chambers and four valves. The left pump is responsible for moving blood through the systemic circulation, while the right pump is in charge of moving blood through the pulmonary circulation. Each pump is made up of a atrium and a ventricle.

The proper functioning of the heart is based in the accurate conduction of electrical impulses that allow for precise and rhythmical atrioventricular contractions. The impulse normally starts at the SA (sinoatrial) node where specialized pacemaker cells send impulses at regular intervals without any type of nerve stimulation. The impulse then travels through the muscles that make up the atria. The interatrial bundle speeds up this process to the left atrium. The atria then begin to contract and the action potential is transferred to the AV (atrioventricular) node via intermodal nodes. There, the impulse slows down thus providing enough time for the atria to complete their contraction. At this point the impulse picks up speed through the bundle of His into the ventricles, through the bundle branches and the purkinje fibers, and both ventricles contract a about the same time.

ECGs are graphs representing the heart's activity based on its impulses.

Figure 1. ECG showing (from left to right) P, QRS, and T waves.

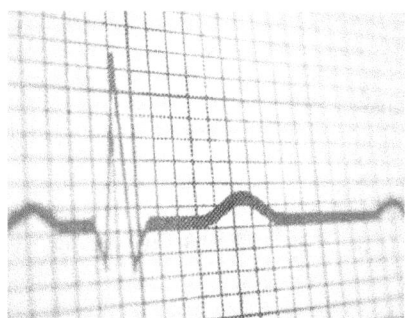

The P wave (first rise on the graph) represents the depolarization of the atria as the impulse goes from the SA node through the atria.

The QRS complex (the highest spike on the graph) represents the depolarization of the ventricles. Voltage fluctuations associated with the atrial repolarization that occur at the same time are too small to be detected here.

The T wave (the short spike that follows the QRS complex on the graph) represents the repolarization of the ventricles.

The cardiac cycle work in two stages: the systole where the heart muscles contract and the diastole where the muscles relax. The cardiac output is the amount of blood ejected by the heart in one minute. When more oxygen is needed in the body, the heart starts beating faster, a response triggered by the sympathetic nervous system.

The systemic circulation is responsible for taking oxygenated blood from the heart to the body and returning deoxygenated blood back to the heart. It occurs through the arteries that bring blood at higher pressure to the organs, arterioles which have muscular walls that allow for the appropriate blood flow to the smaller vessels, capillaries that are responsible for the exchange for nutrients between blood and surrounding tissue, veins and venules that return blood from the capillary bed to the heart. Both the viscosity of the blood and the radius of the vessel it is in affect the rate of flow of the blood.

Pulmonary circulation is responsible for bringing deoxygenated blood back to the lungs and returning oxygenated blood back to the heart. Deoxygenated blood leaves the right ventricle of the heart via the pulmonary arteries, then enter the lungs, and oxygenated blood returns to the left atrium of the heart via the pulmonary veins.

-     Overview of oriental cardio functions

In this section we will discuss the heart and its functions as they are described in traditional Oriental medicine.

The heart is considered to be the most important organ of the human body. Its functions are to govern the blood, control blood vessels and sweat, house the mind. It also manifests in the complexion, and opens into the tongue.

The heart controls the blood by transforming Food Qi into blood. It is also responsible for blood circulation in the body.

It controls the blood vessels by providing enough Qi to have a strong and regular pulse.

The heart controls sweat as blood and body fluids interchange. For example when the blood is too thick, body fluids enter the blood vessels to make the blood thinner. Similarly, profuse and uninterrupted sweating will deplete body fluids and thus the blood as well.

The heart is said to house the mind or Shen. As such it is responsible for the emotional, mental and spiritual balance of the individual. Consciousness, memory, thinking and sleep are also affected by the state of the heart. A deficient heart will not have enough strength to root the mind and will thus give rise to problems such as mental restlessness, depression, anxiety or insomnia, while emotional problems will deplete the heart and cause symptoms such as palpitations.

The heart manifests in the complexion. As the blood is distributed throughout the body, its state can best be seen on the complexion where the blood is closest to the surface.

The heart is responsible for the color, shape and appearance of the tongue, and more particularly the tip. But the heart is also responsible for speech problems such as stuttering. The heart also affects the amount of talking and laughing being done. As such it is said to open into the tongue.

# 14

## Cordyceps and Arrhythmia

### A. Disease mechanism

- Western approach

According to Simon Becker, Bob Flaws and Robert Casanas MD, cardiac arrhythmias refer to any irregular heart rhythm caused by physiological or pathological disturbances in the release of impulses from the SA node or their transmission through the conductive heart tissue.

Five main types are considered in medical practice, the most common being the premature beat. This is a benign condition that may be induced by caffeine, decongestants or stress. Electrolyte imbalances should also be considered when looking at this type of arrhythmia.

Atrial fibrillation is another type, and is the most common of the problematic arrhythmias. Here the two atria quiver unpredictably instead of pumping properly. When this occurs, not all the blood is ejected from the atria. Some remains and may clot. If they break off from the wall of the atria, they may cause a stroke.

Bradycardia refers to the slowing down of the heart to a rate of below 60 beats per minute. Fatigue and fainting spells may occur along with possible dizziness and other neurological manifestations.

Tachycardia refers to a rapid heartbeat which can possibly cause poor blood circulation. A heart rate of above 100 beats per minute is necessary for such as diagnosis.

Ventricular arrhythmias are the most severe of the arrhythmias. They are also the only life threatening ones. They include ventricular tachycardia, and ventricular fibrillation. When ventricular fibrillation occurs a normal rhythm must be reestablished within 3 to 5 minutes brain and heart damage may be sustained and death may result.

They can be related to SA node arrest, abnormal initiation of reentry, or any other abnormal impulse conduction that can occur at any point during the impulse travel through the heart.

- TCM approach

Heart palpitations may occur because of the six external causes, internal damage due to the seven emotions or irregular exercise, poor eating habits, alcohol consumption, poisoning, aging, disease, or other injury.

When the heart is malnourished, not enough Qi and Yang can move the blood or not enough blood and Yin are available to nourish the heart. As such the heart cannot function properly and bradycardia may result.. When the spleen is deficient, it will not be able to provide the heart with the proper nutrients thus leading to heart Qi deficiency and palpitations. When the kidneys are Yin or Yang deficient, they may engender a deficiency of Yin and Yang in the heart as the kidneys are directly linked to the heart. Furthermore any development of lung deficiency will trigger Heart Qi deficiency as well since they both form the chest Qi.

Heat may also rise up to disturb the heart. When this occurs tachycardia may develop. Heat can come from excess or deficiency. Types of heat that generally cause palpitation are deficient heat, phlegm heat, heat turning into stasis, and depressive heat (from Qi stagnation).

Heart Qi obstruction may also cause palpitations. When there is a blockage in the blood flow or phlegm rheum obstruction, the free flow of Qi in the heart is affected and palpitations develop.

## B.    Effects of Cordyceps on arrhythmia

Cordyceps sinensis seems to be quite effective in treating arrhythmias. While its effects are not yet well understood, they may be exerted on the cell membranes by decreasing its permeability to sodium and potassium ions, reducing the repolarization rate and reducing the free radicals available within the cell.

## C.    Research studies on the subject

In 1992, Jingxian Wang published a study on the effect of Cordyceps sinensis capsules on arrhythmias. 52 patients suffering from arrhythmia were given 6 Cordyceps sinensis capsules every 8 hours for 2 months. No other arrhythmia mediation was given. When evaluated, a judgment of excellent meant that there was great improvement of the clinical symptoms and physical signs with the disappearance of premature contraction on the cardiograph. Effective meant that there was improvement of the clinical symptoms and physical signs, with a reduction of premature contraction on the cardiograph. Null meant that there was no improvement of the clinical symptoms and physical signs with no change of premature contraction on the cardiograph. The overall efficacy is the combination of the excellent judgments and the effective ones. The results showed an overall efficacy of 68% and excellent outcome in 47.36% for patients suffering from chest distress. For chest pain, the overall efficacy was 75%, and 59.38% showed excellent results. For premature atrial contractions the overall efficacy was

78%, and 71.43% showed excellent results. For premature ventricular contractions the overall efficacy was 85% and 76.92% of the patients showed excellent results. For sinus tachycardia, the 2 patients suffering from it had excellent results. For sinus bradycardia the overall efficacy was 77% and 55.55% of the patients had excellent results. Improvements were seen between 3 and 20 days, but 69% could be observed after 7 days. No side effects were recorded.

Conclusion: The author concluded that Cordyceps sinensis has a good curative effect on premature ventricular and atrial contractions and that because no side effects were reported that it could easily be used on the elderly.

In 2001 Xiaojian Gong and his Colleagues published a study on the effect of Cordyceps sinensis on arrhythmia. Arrhythmias were triggered in rats using aconitine and barium chloride. Petroleum ether (PE), ethanol (EE), ethyl acetate (EA), water (WE) extract of Cordyceps sinensis, and Mexiletine (Mex) were used to treat the arrhythmias. Extracts were given in different doses in c/mg. kg$^{-1}$ while the duration time of an arrhythmia attacks is in minutes. Duration times were as follows for aconitine induced arrhythmias: control 40.8, PE 400 11.8, PE 200 19.8, EE1500 26.5, EE750 31.3, EA300 29.7, EA150 34.2, WE1500 25.2, WE750 33.8, and Mex60 12.2. Duration times were as follows for barium chloride induced arrhythmias: control 23.7, PE400 15.0, PE200 20.8, Mex60 9.3.

Conclusion: The authors concluded since aconitine speeds up the internal flow of sodium ions, promotes depolarization rate of the membrane, speeds up the automaticity of the pacemaker and shortens the refractory period that PE must be effective in reducing the flow of sodium ions as it showed such improved results. They also concluded since barium chloride promotes the flow of sodium ions in the purkinje fibers, and raises depolarization rate in diastolic period, while it may also decrease membrane permeability to potassium ions, reduce repolarization rate and cause polyphyletic ectopic rhythm, that PE must have an effect on the ion transport through the membrane as it was also effective in reducing arrhythmias.

In 1999 Zhou Yu and Jianxin He published a study on the effects of Cordyceps sinensis on triggered arrhythmias. Cesium chloride was used on rabbits to trigger the response. They measured the monophasic action potential amplitude (MAPA) in mV, the monophasic action potential duration (MAPD90) in ms, The QT in ms, the cycle length (CL) in ms, and the blood pressure (BP) in mmHg. Before the administration of CsCl and Cordyceps sinensis MAPA was 21.2, MAPD90 was 145.5, QT was 201.5, CL was 389.3 and BP was 135.5. After the administration of CsCl MAPA was 22.1, MAPD90 was 252.2, QT was 315.2, CL was 569.2 and BP was 130.0. After the administration of the Cordyceps sinensis powder the MAPA was 20.1, the MAPD90 was 165.2, QT was 288.5, CL was 420.0 and BP was 106.0. Before administration no EAD (early after depolarization) was observed. However, after the administration of the CsCl, high membrane EAD was observed in all the rabbits with an average amplitude of 11.5mV. When Cordyceps sinensis was given to the testees only 3 cases still had EAD with an average amplitude of only 5.2mV.

Conclusion: The authors concluded that the efficacy of Cordyceps sinensis on arrhythmias had been proven beyond a doubt and that the specific mechanism by which it is attained may involve a stable membrane, reduction of $Ca^{2+}$ overload and the reduction of free radicals in cells.

In the last couple of decades, Cordyceps sinensis studies have shown that the fungus in indeed effective at treating arrhythmias even if the mechanism of action is not completely understood.

# 15

## Cordyceps and Pulmonary Heart disease

### A.    Disease mechanism

-       Western approach

When blood oxygen levels remain low for too long, the body compensates by increasing the pulmonary arterial pressure. When this occurs for some time, Cor pulmonale develops.

Cor pulmonale (or CP) is right sided heart failure triggered by long standing lung disease. Indeed the right ventricle becomes enlarged as the pressure increases in the arteries of the lungs.

As the right ventricle starts failing and the intrathoracic pressure increases, distention of the veins and peripheral edema develop.

COPD, loss of lung tissue such as surgery or other trauma, pulmonary emboli, scleroderma, neuromuscular diseases affecting the muscles involved in respiration and other causes resulting in alveolar hypoventilation may cause CP. It is also often associated with recurring pulmonary infections.

Hypercapnic acidosis low oxygen levels in the blood are hallmarks of this disease. X-rays and other imaging techniques showing an enlarged right ventricle and proximal pulmonary artery are helpful tools in establishing a diagnosis of CP.

- TCM Approach

According to traditional Oriental medicine CP stems from a deficient root and an excess branch.

Recurring illnesses affecting the lungs will weaken the organ's Qi thus leading to a deficient protective Qi and a poorly protected surface. Exterior pathogens will then have an easier time at invading the lungs and causing coughing and sputum evacuation.

When chronic lung disease is involved, the spleen becomes implicated as well. A deficient spleen will not be able to metabolize water properly and dampness and phlegm will be produced. They will then accumulate in the lungs. Furthermore the inability of the deficient spleen and lung to move blood will create blood stasis and thus lip cyanosis and palpitations associated with heart problems. Deficient spleen and lungs will also affect the kidney thus leading to an inability to receive Qi and the resulting wheezing. A deficient kidney will also cause further fluid metabolism problems with the formation of edema, and scanty urination. As the water accumulates, it eventually flows to the heart causing shortness of breath and palpitations.

## B.  Effects of Cordyceps on pulmonary heart disease

Cordyceps sinensis has been shown to decrease the amount of time spent in the hospital by patients suffering the chronic pulmonary heart disease. It has also been shown to decrease the clinical manifestations of the disease and reduce the necessity for respirators.

It has also been shown to affect amino acid reserves in such patients and thus reduce the impact they have on respiration.

# C.   Research studies on the subject

In 2002, Zhibing Yang and his colleagues studies the effects of Cordyceps sinensis on 30 Cor pulmonale patients. Patients were divided into a CS (Cordyceps sinensis) group that contained 30 individuals and a control group that contained 20. The CS group was given 10ml of liquid Cordyceps sinensis (the equivalent of 5g of Cordyceps sinensis) twice daily for 2 weeks. The control group received standard drug therapy. Hospitalization time was 2 to 3 weeks with no use of respirator for the CS group while the control group's hospitalization was 4 to 6 weeks with 3 cases using respirators.

Serum amino acid levels were also observed and recorded in µmol/L. Branched chain amino acids (BCAA) include leucine, isoleucine and valine. The aromatic amino acids (AAA) include phenylalanine, histidine, and tyrosine.  For Leucine, in the CS group, levels were 120.92 before treatment and 122.2 afterward. For the control group it was 115.2 before treatment and 117.26 afterward. For Isoleucine, in the CS group, the levels were 45.76 before treatment and 68.5 afterward. For the control group it was 48.15 before treatment and 58.76 after. For Valine, in the CS group, levels were 178.52 before treatment and 217.11 after. For the control group it was 189.86 before and 194.52 after. For Tyrosine, in the CS group, the levels were 80.26 before treatment and 57.51 afterward. For the control group it was 81.76 before treatment and 72.5 after. For Phenylalanine, in the CS group, the levels were 90.22 before treatment and 72.15 afterward. For the control group it was 94.5 before treatment and 98.22 after. For Histidine, in the CS group, the levels were 72.56 before treatment and 85.46 afterward. For the control group it was 72.66 before treatment and 82.40 after.

The ratio of BCAA/AAA was 2.00 before treatment for the CS group and 3.10 after, while the before treatment ratio was 2.00 and the after treatment was 2.10 for the control group. Overall BCAA decrease and AAA increase in chronic respiratory failure patients. Phenylalanine and tyrosine are of particular interest as they are precursors of catecholamine. An excessive amount of these amino acids will cause the dehydroxylation of tyrosine to create tyramine. This false neurotransmitter will then compete with dopamine and make respiratory failure more severe.

Conclusion: The authors concluded that Cordyceps sinensis, by increasing the BCAA/AAA ratio, decreased the number of false neurotransmitters and thus improved respiration. As such they further concluded that the fungus was effective in treating pulmonary heart disease and respiratory failure.

In 2006 Lang Xiao and his colleagues published a study on the effect of Cordyceps sinensis and Corbrin capsule on chronic pulmonary heart disease with respiratory failure patients. 80 individuals were divided into 3 groups: control group, Cordyceps sinensis (CS) group, and Corbrin group. All group received standard treatments, but the CS group received 10ml of Cordyceps sinensis (5g equivalent) twice daily and the Corbrin group received 5 capsules 3 times a day of Corbrin. The results for the CS group were exactly the same as the previously described study and the Corbrin group exhibited similar results. It is strongly suspected that the author used his previously published work and added the Corbrin study to it.

Conclusion: The authors concluded that both Cordyceps and Corbrin were good adjunct treatment for patients suffering from chronic pulmonary heart disease with respiratory failure.

Cordyceps sinensis does indeed appear to help with the sign and symptoms associated with pulmonary heart failure.

# 16

## Cordyceps and Chronic heart failure

### A.    Disease mechanism

-       Western approach

Chronic heart failure, or congestive heart failure, is the dysfunction of the cardiac muscle caused by a structural or functional problem that reduces the ability of the heart to move the blood through the body effectively.

It is generally caused by a previously developed illness and may involve either ventricle.

In left ventricular failure, cardiac output is reduced and the pulmonary venous pressure in increased. The lungs get filled with water and slowly lose their ability to stretch, while breathing in oxygen becomes more taxing. Indeed when the hydrostatic pressure in the veins of the lungs surpasses that of the plasma protein osmotic pressure, fluids start leaking into the capillaries, interstitial space and alveoli thus causing a reduction in alveolar ventilation.

In right ventricular failure, the body as a whole is generally affected and edema of the lower extremities develop. Liver functions also become slightly compromised and cellular death may also be observed around the hepatic vein.

Systolic and diastolic pressure may become abnormal. During systole, the ventricle may not empty completely thus increasing the pressure stretching the ventricle as well as the diastolic volume. Such stress will lead to premature aging and death of cardiac muscle cells and an abnormal enlargement of the ventricles. As the amount of blood pumped out of the ventricles with each heart decreases, the heart begins to fail. When diastole becomes abnormal, filling of the left ventricle becomes more difficult, the ventricles relax for longer periods of time and ventricular filling is affected as well. In this instance, the amount of blood pumped out of the heart may increase or remain the same.

Regardless of what causes chronic congestive heart failure, blood circulation, kidney function, and hormones will be affected. Muscle cells may be lost, excess potassium may be secreted, arrhythmias may develop, glomerular filtrate rate may decrease while tubular reabsorption may increase. The rennin-angiotensin-aldosterone system becomes deregulated and water retention results.

- TCM approach

Traditional Oriental medicine considers CHF to be caused by congenital causes, irregular exercise, eating and drinking, damage by the seven emotions, long standing chronic disease and/or aging.

Four organs appear to be responsible for CHF: the heart, the lungs, the spleen and the kidney. The heart moves the blood through the body with the help of the ancestral Qi that is generated by the lung. So when the lungs become weak, the heart can no longer function properly either. The spleen moves and transforms fluids, and produces blood. When the spleen is weak, blood is not produced effectively and the heart is affected. The kidneys support the spleen by providing it with Yang energy, and transforms water. When water is not metabolized properly it floods up to the heart to damage heart yang and cause disruptions in the blood flow. Water accumulation will manifest under the skin and may even collect in the chest, face and/or legs.

## B.    Effects of Cordyceps on chronic heart failure

Cordyceps sinensis does affect the quality of life of patients with congestive heart failure. Not only does it measurably decrease their shortness of breath, fatigue, depression and anxiety, but it also increases their stamina, self control and general sense of well being.

Some scholars suggest that the fungus can expand coronary vessels, increase blood flow to cardiac muscle, and increase hypoxia tolerance in atmospheric pressure.

On the more objective note, both cardiac output and stroke volume appear to be positively affected by Cordyceps sinensis.

## C.    Research studies on the subject

According to Georges Halpern MD, a study was published in 1995 by researcher from Fu-Jian Medical College. 64 patients were divided into a control group of 30 individual and a treatment group of 34 people. Both group received conventional western medicine, and the treatment group was given 3 to 4 grams of Cordyceps sinensis daily. 66% of the patients on Cordyceps suffered less from shortness of breath and fatigue after their treatment. Only 25% of the patients in the control group reported the same thing. Progress was also observed when it came to cardiac output, stroke volume and heart beat. For physical activity, the control group showed improvement in 40% of the cases while 79.41% of the treatment group exhibited similar improvement. Increased sex drive, well being, and self control along with a decrease in anxiety and depression were also observed in patients taking the fungus.

Conclusion: The authors then concluded that Cordyceps sinensis was indeed an appropriate treatment for patients suffering from Congestive Heart Failure.

In 1995 Xuekui Gu and his colleagues published a study on the effect of ginseng Cordyceps sinensis soup on chronic congestive heart failure. 32 patients were included in the study. All were to continue necessary western treatments such as cardiac glycosides, vasodilators and diuretics. They were also given preparations of 15g ginseng and 10g Cordyceps sinensis decocted for 2 hours in 150mL of water to be taken for a period of 2 weeks. Patients were evaluated for improvement using the NYHA classification. If they improved by two levels during the course of treatment they were considered to have had a very effective treatment. If they improved by one level the treatment was classified as effective. The treatment was also deemed ineffective if the patient did not change levels what so ever. Following these guidelines, 3 cases showed no improvement, and 23 cases showed treatment effectiveness. The decoction proved to be very effective in 6 cases.

Conclusion: The authors concluded that the combination Ginseng Cordyceps sinensis offered good support for congestive heart failure patients.

These studies seem to show that indeed Cordyceps sinensis is indicated for the treatment of chronic heart failure. The physical improvements associated with the consumption of the fungus appear to warrant its use for this disease as it not only improves the health of the individual taking it but also the general quality of life of the patient.

# 17

## Cordyceps and atherosclerosis

### A.  Disease mechanism

-   Western approach

Atherosclerosis is a disease that affects medium and large sized arteries such as the aorta, the coronary arteries, or those irrigating the brain. It is characterized by the formation of plaques or atheromata. It is the most common type of arteriosclerosis.

During the development of this disease, there is an accumulation of lipids on the lining of the arteries, with an increase in blood vessel smooth muscle cells and the creation to scar tissue. A grey elevated thickening with a lipid core and a fibrous covering of connective tissue and smooth muscle develops on the vessel wall. Eventually scar tissue, ulcerations and hemorrhage may be observed. The lesion grows and effectively reduces the diameter of the blood vessel, thus reducing the flow of blood going through it.

Injury to the wall of the blood vessel appears to be the trigger for the initial inflammatory response that leads to atherosclerosis. During this process, white blood cells, platelets, cholesterol and various blood proteins increase the formation of smooth muscle cells and connective tissue within the blood vessel wall. Factors such as smoking, mechanical stress caused by hypertension or immune responses may play a role in the development of this disease.

Diseases such as Coronary Heart disease are included in the illness category.

## - TCM approach

According to traditional Oriental medicine atherosclerosis can be caused by poor nutrition and drinks with irregular schedules, over work or exercise, excessive sexual activity, poor bowel habit or uncontrolled emotions.

Atherosclerosis may be derived from internal heat, lack of fluids on the blood vessel walls which will lead to heat, and/or the accumulation of dampness in the blood in an effort to clear the heat.

Generally heat first accumulates in the liver as anger and other emotions are repressed and liver Qi stagnates for some time. Since the liver stores the blood, the heat present in the liver will progressively transfer to the blood and then to the blood vessels as the blood travels through the body. This is viewed by western medicine as the inflammation process.

The hardening and loss of elasticity of the blood vessels is linked to the presence of heat as well. The body will indeed try to compensate by bring cooling fluids from the kidneys to the liver and blood. Eventually, fluids become depleted and the body dries up.

## B.    Effects of Cordyceps on atherosclerosis

Cordyceps sinensis, when associated with atherosclerosis, appears to increase the immune response of red blood cells by increasing their activity in removing circulating immune complexes from the walls of the arteries thus reducing the inflammatory response. It may also reduce the cholesterol levels present in the blood, and improve the activity of phagocytic white blood cells.

## C.    Research studies on the subject

In 2000 Yu Yamaguchi and his colleagues published a research on the inhibitory effects of Cordyceps sinensis water extract on raised serum lipid peroxide levels and aortic cholesterol deposition in atherosclerotic mice. Groups of 7 mice each were established. The normal group was fed a basal diet. The other four groups were fed atherosclerosis producing diets. The control group was fed only that while the WECS50 group was also fed 50mg/kg/day of Cordyceps sinensis extract, the WECS100 received 100mg/kg/day for the extract and the WECS200 received 200mg/kg/day of the fungus extract. All groups received their prescribed regimen for 12 weeks. Weight gain was observed in all groups with no significant variation. The normal group total cholesterol (in mg/100mL) was 196, free cholesterol (in mg/mL) was 44.3, phospholipids level (in mg/mL) was 265 and triglycerides (in mg/100mL) were 162. The control group total cholesterol (in mg/100mL) was 412, free cholesterol (in mg/mL) was 93.4, phospholipids level (in mg/mL) was 330 and triglycerides (in mg/100mL) were 102. The WECS50 group total cholesterol (in mg/100mL) was 310, free cholesterol (in mg/mL) was 70.3, phospholipids level (in mg/mL) was 264 and triglycerides (in mg/100mL) were 109. The WECS100 group total cholesterol (in mg/100mL) was 385, free cholesterol (in mg/mL) was 88.3, phospholipids level (in mg/mL) was 272 and triglycerides (in mg/100mL) were 106. The

WECS200 group total cholesterol (in mg/100mL) was 495, free cholesterol (in mg/mL) was 112.4, phospholipids level (in mg/mL) was 328 and triglycerides (in mg/100mL) were 105. Aortic cholesterol levels were also obtained. At 12 weeks the normal group showed levels (in mol/mL) of about 4, the control of about 9, the WECS50 of about 6.5, the WECS100 of about 6 and the WECS200 of about 5.5.

Conclusion: The authors concluded that the Cordyceps sinensis can reduce lipid peroxidation in LDL by collecting and removing free radicals thus hindering cholesterol build up on the aorta's wall.

In 2005 Xiao Ma and his colleagues published a study on the effect of Cordyceps sinensis on atherosclerosis. For this experiment they fed high fat diets to 20 rabbits for 12 weeks. Cordyceps sinensis was added to their diet for the last 8 weeks before testing. Results were as follows. For CH (high fat cholesterol) the basic status was 1.46, after high fat diet 22.66, with Cordyceps sinensis 10.26. For TG (triglycerides) the basic status was 1.08, after the high fat diet 2.18 and after the Cordyceps sinensis 1.21. For HDL-c the basic status was 0.37, after the high fat diet 0.33 and after Cordyceps sinensis 0.39. For LDL-c the basis status was 0.63, after the high fat diet 21.35 and after the Cordyceps 9.32. All prior results were expressed in $c/mmol.L^{-1}$. The researchers also checked the percentage rosette formation rate and the circulating immune complexes present in the serum of the rabbits. For RBC-$C_3$bRR (rosette rate of red blood cell C3b receptor) the basic status was 10.36, after the high fat diet it was 6.44 and after intake of Cordyceps sinensis it was 9.77. For RBC-ICR (rosette rate of red blood cell immune complexes) the basic status was 13.25, after the high fat diet it was 20.27 and after intake of Cordyceps sinensis it was 15.21. For CIC (Circulating immune complexes) the basic status was 156.60, after the high fat diet it was 537.55 and after intake of Cordyceps sinensis it was 233.45. IMTm (maximum intimal medial thickness) of the abdominal aorta was 0.32mm before ingestion of the high fat diet and 0.89mm after. After the Cordyceps treatment it was down to 0.52mm.

Conclusion: The authors concluded that Cordyceps sinensis helps reduce the negative impact of a high fat diet on the body. The fungus appears to increase the ability of red blood cells to clean CIC deposit on arterial walls by increasing the activity of receptor $C_3b$ present on the red blood cells membrane. This will then decrease the IMT of the arteries. In their words:" To put it together, reduction of RBC immunity is one of the most dangerous factors in atherosclerosis and Cordyceps sinensis compound may be a promising anti-atherosclerosis medicine that is worthy of further research".

It is interesting to notice that in spite of increasing cholesterol levels with increasing dosage of Cordyceps that the aortic lipid deposits pattern was going in the reverse direction. One may be tempted to say that the most important factor in the development of atherosclerosis may indeed be the inflammation process and not as much the cholesterol levels. But regardless of the mechanism, Cordyceps has been shown in these studies to be quite effective in controlling the development of atherosclerosis.

# Part F

# Cordyceps

# And

# Other uses

# 18

## Cordyceps and Fatigue

## A.  Disease mechanism

-  Western approach

Chronic fatigue syndrome (CFS) is a disease that may affect the body in its entirety. It is characterized by a debilitating fatigue that lasts at least 6 months, that is not relieved by bed rest, and that may follow a period of flulike symptoms. It may include unwarranted fatigue after normal activities, reduced ability to maintain energy levels during a job that requires continued attention, or a general inability for action. Sore throat, lymph node pain and tenderness, headache, impairment of memory or concentration, muscle pain and joint pain may also be present.

The exact pathophysiology of this disease is mostly unknown. Theories have been put forth as to the involvement of infectious agents such as EBV, Mycobacterium tuberculosis and Candida.  Possible genetic predisposition may also be involved. According to Carol Mattson Porth 40% to 50% of CFS sufferers had had depression prior to coming down with the disease. Also some uncharacteristic immunological responses have been observed in these patients thus suggesting an overreaction to an environmental agent or an internal stimulus that would trigger an inability to self regulate once the infection is over.

- TCM approach

According to traditional Oriental medicine the development of CFS is based on the dysfunction of a few organs, the most important of which is the liver. This organ is indeed the origin of extreme fatigue. It normally dominates the sinews. But when it stops doing so, the patient will develop an aversion to activity thus leading to fatigue.

Furthermore as the internal organs are interdependent, the liver will affect the lungs, the heart, the spleen, and the kidneys. So ultimately, emotional stress, which would act on the liver, can also act on the other organs. Any other activity that will tax Essence and Qi will lead to consumptive deficiency, the logical progression of CFS.

Motor function problems associated with CFS are caused by dysfunctions of the liver, spleen and kidneys. The liver dominates the sinews. When liver blood becomes deficient, the sinews are malnourished thus leading to lack of strength. The spleen dominates the muscles. When the spleen is deficient and both blood and Qi are deficient as well the muscles will become sore and weak. When dampness develops, heaviness and weakness in the extremities increase as well. The kidney dominates the bones. Kidney essence deficiency will lead to low back weakness and soreness and the inability to stand or walk. When the marrow is no longer nourishing the bones, sore, red, swollen, and painful joints may develop.

Memory problems are associated with brain impairment. The kidney stores essence which in turn generates marrow which then nourishes the brain. As the essence becomes depleted, the marrow becomes depleted as well and the brain is not nourished properly. The spleen is also involved with memory. Liver Qi stagnation and spleen deficiency leading to phlegm damp accumulation are related to brain function. Phlegm damp misting the orifice will indeed trigger impaired memory.

## B.   Effects of Cordyceps on fatigue

Authors speculate on the efficacy of Cordyceps sinensis in the treatment of this disease based on the inspection of various clinical observations. CFS patients develop a specific type of adrenal deficiency along with high levels of testosterone. They also exhibit abnormal respiratory muscle performance and hypothalamic-pituitary-adrenal axis dysfunctions that lead to depression.

Since CFS may be linked to an immune response gone wild, Cordyceps sinensis may be helping with regulating the immune system and thus controlling the development of this disease.

Cordyceps is also known for its ability to reduce fatigue, the hallmark symptom of chronic fatigue syndrome, by boosting their energy.

## C.   Research studies on the subject

Precious little is available in terms of research when it comes to Chronic fatigue syndrome. Because the disease mechanism is for the most part unknown very little research has been done on the subject.

However a research studying the effect of Cordyceps sinensis on the aerobic capacity and respiratory function of elderly healthy individuals was published in 2004. 37 subjects were placed in a treatment and a control group. For 6 weeks, the treatment group was given 3g of Cordyceps sinensis per day in capsule form while the control group was given identical placebo capsules. Maximum oxygen uptake (in L/mn) was 1.88 before treatment and 2.00L after treatment for the treatment group, and 1.80 before treatment and 1.79 after for the control group. Maximum ventilation (in L/mn) was 52.1 before treatment and  51.3 after for the control group, and 56.7 before treatment and 60.4 after for the treatment group. Maximum work rate (in watt) was 123.2 before treatment and 117.9 after for the control group, and 123.4 before treatment and 128.0 after for the treatment group.

Conclusion: The researchers concluded that Cordyceps sinensis not only improved the capacity of the subject to exercise but also increased their resistance to fatigue.

The limited study on fatigue and the previously discussed ones in the immunity section may lead us to suggest that Cordyceps sinensis may be beneficial to the treatment of chronic fatigue syndrome. However, further studies would be required for a more definite endorsement of this therapy for this particular ailment.

Note: There appears to be one consistent finding in my own practice. Patients placed on Cordyceps for whatever reason, seem to all feel a boost in energy, less fatigue, and as one of my favorite patients would say, generally more "umph".

# 19

## Cordyceps as a sexual tonic

### A.  Overview of sexual dysfunctions

Sexual dysfunctions are problems that hinder the initiation, consummation, or satisfaction with sex. They happen in both men and women and are not linked to sexual orientation. They can be lifelong, acquired, situational or generalized.

Of the more common ones are seminal emissions, erectile dysfunction (ED), and Inhibited sexual desire (ISD). Infertility is not a sexual dysfunction but is related to the topic as it is often caused by an underlying kidneys deficiency.

Seminal emission, or spermatorrhea, refers to the unprompted discharge of semen without sexual activity. It may be nocturnal during sleep or during the day while awake. It is due to kidney Qi failing to control Essence. When the kidney Qi is weak, the stored Essence leaks out.

Associated with this disease are four main patterns that may be responsible for kidney Qi deficiency: heart and kidney disharmony, damp-heat in the lower Jiao, deficient Qi not being able to control essence, and kidney deficiency with seminal gate instability.

Heart and kidney disharmony can lead to nocturnal emission and is the most common pattern when it comes to this disease. Heart fire becomes overactive when the heart yin is depleted because of excessive intellectual work. The heart fire then disturbs the mind and causes excessive dreaming with nocturnal emission. Furthermore, the overactive heart fire and the yin deficiency will also stir up the kidney and agitate the seminal vesicles, thus causing nocturnal emission.

When there is excessive eating of greasy rich foods and excessive drinking of alcohol, the spleen becomes deficient. The damp-heat that results is sent down to disturb the seminal vesicles and thus creating sperm leakage day and night.

Physical and mental overwork can lead to heart and spleen Qi deficiency. When the deficient Qi does not succeed in controlling essence seminal emission will occur.

Kidney yin deficiency with empty fire will cause a disruption of the seminal vesicles, thus leading to nocturnal emission. When this condition is long lasting, both essence and Qi deficiency will develop. This will complicate the issue further and cause wakeful seminal emission as well.

Erectile dysfunction, or impotence, refers to the recurring or constant inability to attain or maintain an erection. Causes of ED include drugs, alcohol diabetes, spinal cord injury and other illnesses. Western medicine suggests the infamous blue pill (Viagra) as a treatment for this problem. Traditional Oriental medicine considers that it may be caused by Kidney Yang or Ming men fire deficiency. Heart and spleen deficiency due to excessive thinking and worry may also cause Qi and blood deficiency leading in turn to erectile dysfunction. Fear is the emotion of the kidney. As such, an excess of fear will damage the kidney, dispersing essence and causing kidney dysfunction and thus impotence in the process.

Inhibited sexual desire, or lack of libido, is characterized by decreased sexual attraction and sexual activity, limited or nonexistent sexual dreams, fantasies, or interest in erotic material. Its causes may include hormone imbalances, depression, alcohol, liver or kidney disorders, chronic illnesses, drugs (prescribed or otherwise), stress, or sexual trauma. In traditional Oriental medicine it is directly linked to a kidney yang and/or essence deficiency.

Infertility is defined in western medicine as the inability to achieve pregnancy after one year of normal sexual activity without the use of contraception. According to western medicine, female infertility may be caused by ovulation factors, tubal factors, or cervical factors. Treatments for cervical factors include conjugated estrogens or intrauterine insemination with the partner's sperm.

Treatments for tubal and uterine factors generally involve surgery. In Vitro fertilization may be used when tubal damage is too severe. It is also used regularly with endometriosis, the presence of sperm antibody and other undiagnosed causes. IVF is performed by triggering ovarian hyperstimulation, followed by oocyte retrieval, then fertilization and embryo culture, and finally embryo transfer into the uterus.

According to Jane Lyttleton, TCM looks at ovulation and implantation specifically when determining the cause of female infertility. Ovulation problems are characterized by amenorrhea, very irregular cycles, a basal body temperature chart that shows no change in temperature or when blood tests indicate low estrogen and/or progesterone levels or high FSH levels. Poor ovulation can also be suspected when the basal body temperature chart indicates only a marginal rise in temperature or a very short lived rise, when blood tests indicate low levels of progesterone at mid-cycle, or when there is spotting in the luteal phase. Problems with implantation will occur when there is poor ovulation, when fibroids or other physical defects are present in the uterus, when the lining of the uterus is deficient, when there is a very little, dark or clotty menstrual flow, or when all else has been ruled out and no other explanation is plausible.

Problems related to ovulation can be linked to Kidney Yin Xu (most common), or Heart Qi stagnation with possible complication of Liver Qi stagnation, phlegm-damp or blood stasis in either case. When primary amenorrhea is involved a Chong and Ren Mai imbalance is often responsible. With secondary amenorrhea due to anovulation the Ren, Chong, and Du Mai are involved as well as the Lung and Spleen. Problems with implantation are most often associated with Kidney Yang Xu with spleen deficiency possibly complicated by Liver Qi stagnation, Phlegm-Damp or blood stasis. Kidney Jing Xu is also very often present in all instances mentioned above. Qi and blood imbalance, along with Yin Wei and Ren meridians, Spleen, Kidney and shen problems also affect implantation. Congenital deficiencies also observed in female infertility consist of specific signs that include a spiral shaped fold of the vaginal orifice, a narrow vagina, constriction of the vaginal orifice as if it is absent, and hypertrophy of the clitoris. Cervical mucus problems are caused by Qi and blood imbalance, Spleen and Liver problems as well as Chong Mai and Shen disturbance.

Male infertility is related to the absence of healthy sperm capable of moving all the way up to the ovum in order to fertilize the latter. In Chinese medicine it is directly related to the kidney. It can either be kidney yin, yang, or essence deficiency, or all of the above. Damp-heat and blood and Qi stagnation may also be cause of infertility in males.

Kidney Yang deficiency is the most common imbalance causing infertility in men. Damp-heat is generally associated with some type of infection. Qi and blood stagnation cause obstruction of the sperm passages and can be caused by damp-heat, congenital abnormality, varicoceles, or testicular trauma.

## B.    Effects of Cordyceps on erectile dysfunction

Cordyceps sinensis has been known to the Chinese to alleviate impotence for about two thousand years. It has recently been shown as well in clinical trials, but the mechanism of action is largely still unknown.

## C.    Effects of Cordyceps on infertility

Cordyceps sinensis is known to replenish sperm as it is a kidney tonic and sperm problems are associated with kidney deficiency in traditional Oriental medicine. It was shown in recent clinical studies to increase sperm count, and survival time while decreasing possible abnormalities.

## D.    Effects of Cordyceps on sex drive

Cordyceps has been known to the Chinese as an aphrodisiac for almost two thousand years. Recent studies have also shown a definite increase in libido after the ingestion of Cordyceps sinensis for some time. The mechanism of action appears to be linked to the ability of Cordyceps sinensis to induce sex hormone like responses in the patients. It has also been suggested that Cordyceps sinensis could affect the hypothalamo-pituitary-adrenocortical axis thus affecting the sexual center of the brain.

## E.    Research studies on the subject

In 2008 Yuan Hongfen and Zhu Jianping published a study on the effect of Cordyceps sinensis in the treatment of seminal emission. 15 patients were part of this experiment. Chickens were cooked with 25 to 30g of Cordyceps sinensis in their abdominal cavity. One chicken was to be eaten in 3 to 4 days and 4 chickens were to be consumed before the patients were re-evaluated. 2 of the patients experienced a complete recovery. 6 patients showed marked improvements, 5 improvement, and 2 no improvement at all. The total effective rate was therefore 86.67%.

Conclusion: Cordyceps sinensis does indeed help in the treatment of seminal emission.


In 2001 Bu-Miin Huang and his colleagues published a study on the effect of Cordyceps sinensis on testosterone production in normal mouse Leydig cells. Cells from the testes of mice were harvested and incubated at 37°C. They were treated with different doses of Cordyceps sinensis and with or without hCG. Testosterone levels were measured in pg/50000cells/3hrs. Cordyceps sinensis (CS) was expressed in mg/mL The control was at about 290, CS 0.1 at about 280, CS 0.3 at about 450, CS 0.6 at about 400, CS 1 at about 620, CS 3 at about 800, CS 6 at about 420 and CS 10 at about 150. When 50ng/mL was added to the Cordyceps sinensis the following results were observed: the control was about 300, hCG only 11200, hCG+CS 0.1 about 12800, hCG+CS 0.3 about 11600, hCG+CS0.6 about 9700, hCG+CS 1 about 8700, hCG+CS 3 about 6000, hCG+CS 6 about 3000, and hC+CS10 about 3200.

Conclusion: The authors concluded that CS could indeed stimulate steroidogenesis but that it was dosage dependant and that once hCG was involved that it inhibited the production of testosterone at high dosage. They suggested that this phenomenon could be explained if one considered that CS affects steroidogenesis by influencing a step in the signal transduction pathway after the formation of cyclic AMP (secondary messenger in the signal transduction cascade).


The following research was reported by Jia-Shi Zhu and his colleagues in their paper titled "The Scientific Rediscovery of an Ancient Chinese Herbal Medicine: Cordyceps sinensis Part I".

In 1995, Yang and his colleagues published a study on the effect of Cordyceps sinensis on libido. Two groups were established, one receiving Cordyceps sinensis while the other received placebo. The Cordyceps group received 3g of Cordyceps sinensis per day for 40 days. For impotence 23% the patients showed improvement with the placebo while 66% of the subjects treated with the Cordyceps sinensis showed similar results. For seminal emission 0% of the patients taking the placebo showed improvement while 67% of the Cordyceps recipients improved. For men libido did not improve for the placebo group, while 67% of the Cordyceps sinensis group showed improvement. For women, libido did not improve for the placebo group, while the Cordyceps group showed improvement in 86% of cases. Other sexual dysfunctions in women showed a 100% improvement when using Cordyceps and none when using the placebo.

Conclusion: Cordyceps sinensis seemed to help considerably with increasing libido in both sexes.

In 1988 Wan and coworkers published a study on the effect of Cordyceps sinensis on decreased libido. The placebo group showed improvement in 23.7% of the cases while the Cordyceps group showed improvement in 66.1% of the cases.

In 1986, Guo and his colleagues published a study on the effect of Cordyceps on impotence and other sexual dysfunctions. Patients were given 3g of Cordyceps sinensis per day for 8 weeks. Before the treatment 0% of the subjects were capable of sexual intercourse, while 37% could perform after treatment. Sperm count (in million/mL) was 99 before treatment and 132 after. Sperm malformation was 70% before treatment and 50% after. Survival rate was 29% before treatment and 52% after.

All these studies seem to indicate that Cordyceps sinensis does indeed have a positive effect on the treatment of sexual dysfunctions. However, the mechanism of action remains vague and further study would be required to get a more thorough understand of the step by which Cordyceps seems to help these various ailments.

# Conclusion

Cordyceps sinensis has been considered a treasure of nature for almost two thousand years in China. It has been used by royalty to treat ailments going from cough and other lung problems to kidney deficiency and bleeding issues. In recent years this amazing fungus has been discovered by the West, and studies were developed in an attempt to validate the claims made by its avid supporters. Illnesses such as diabetes, cancer, AIDS, lupus, chronic and acute renal failure, kidney transplant, fatty liver, cirrhosis, hepatitis, bronchitis COPD, asthma, arrhythmias, pulmonary heart disease, chronic heart failure, atherosclerosis, fatigue, and a number of sexual dysfunction have been help by it.

In general the studies dealing with each illness were quite efficient at demonstrating the ability of Cordyceps sinensis to help with a particular ailment, but were not so effective at establishing the mechanisms of action leading to such results. Because of that the biochemical reasons behind the success of Cordyceps sinensis in the treatment of the above mentioned illnesses remains mostly unknown. Sadly, western practitioners tend to rely solely on that to validate a treatment protocol. So many patients may be left in the dark as to their option to use this fungus (or any other herbal formula) as a treatment or an adjunct to what their practitioner is already prescribing to them.

The use of Cordyceps sinensis in the treatment of diabetes is one of the better studied subjects as of today. It has been shown to lower blood sugar, increase serum insulin levels and possibly reduce insulin resistance. Some have suggested that the fungus may calm the immune system in order to regulate the autoimmune response associated with type 1 diabetes. In my own practice, I have seen it regulate blood sugar levels time and again in even the most difficult of cases, Patients were taken of insulin and other western drugs with the help of this herb.

Cordyceps has shown to be quite effective in the treatment of cancer. T cells and lymphocyte production seems to be boosted by the use of this fungus. CBC tends to remains normal in more patients when Cordyceps sinensis is used in conjunction with chemotherapy. Symptoms associated with the disease improve. Fever is reduced and anorexia dissipates. Patients experience less vomiting, fatigue and pain. Furthermore, the fungus appears to decrease the growth of cancer cells. Some have also suggested that part of of the benefits of using Cordyceps sinensis on cancer patients was that the fungus seemed to accelerate the function of the scavenging tumors cells present in the body and thus played a role in preventing metastasis. However the exact mechanism of action remains mostly unknown.

HIV/AIDS can apparently also be helped with Cordyceps sinensis. The fungus seems to be enhancing CD4 helper T cells. Not only does it seem to increase the T cells proliferation but it also appears to strengthen their function. Regardless of the actual mechanism used by the herb to combat this disease, a strong anti-viral action has been observed. However researchers agree that more study is needed to understand the way Cordyceps sinensis affects the body of those infected with HIV.

The use of the fungus in the treatment of lupus also appears promising. It has been shown to increase the life expectancy of patients suffering from this disease possibly by reducing or even stopping the production of antibody. It has also been shown to inhibit lymphadenectasis, reduce proteinurea, and improve kidney function overall.

Cordyceps sinensis appears to help with renal failure by redistributing epidermal growth factors of renal tissues, accelerating the recovery of renal tubules, reducing the influx of calcium into the renal cortex and the peroxidation of cellular lipids, and improving mitochondrial energy metabolism. Ultimately the fungus promotes healing and protects the kidneys from further damage.

In the case of kidney transplant, Cordyceps sinensis helps by reducing the side effects of drugs such as gentamicin and cyclosporine. Indeed these chemical compounds are known to cause kidney damage, a particularly ironic situation in the case of Cyclosporine as it is given as an anti rejection drug after transplant.

Cordyceps has undeniable protective and reparative effects on the liver. By a complex mechanism it appears to reduce the deposition of collagen in this organ thus slowing down the disease progression. Studies also indicate that the fungus help regulate liver function as it helps normalize liver enzymes.

In a similar manner, Cordyceps sinensis appears to help patients suffering from cirrhosis. It seems to stop hepatocyte degeneration, decrease the collagen amount present in the liver and reduce the size and number of varicosities detected in these patients. It also appears to increase metabolism and ATP production thus promoting cell repair, and decrease the pressure in the hepatic portal vein. Again the mechanisms of action for these processes are still unclear and require more study.

When it comes to hepatitis, Cordyceps sinensis regulates immune responses leading to the search and destruction of infected cells. It also softens the liver by decomposing collagen deposits, promote cellular regeneration, reduces fatigue, normalizes liver enzymes and help alleviate symptoms in general. But even if the chances of recovery of patients suffering from hepatitis are improved by the use of Cordyceps sinensis, the mechanism by which it is achieved remains poorly understood.

Cordyceps sinensis appears to help chronic bronchitis by reducing the inflammation present in the lungs. However patients suffering from concurrent heart disease seem to not reap the benefit of this herb. Further study is definitely required to investigate the impact of the fungus on this disease and more specifically the mechanisms by which it is achieved.

COPD can be helped by Cordyceps sinensis since the fungus inhibits tracheal contraction, has an anti-inflammatory effect, and increases gas transport in the alveoli. It also stops coughing and reduces phlegm. As an added bonus the fungus appears to reduce the hypertension in the pulmonary small vessels thus reducing the risks of developing chronic pulmonary heart disease and right heart failure.

Cordyceps seems to help asthma sufferers by reducing the inflammation partly responsible for obstructing the airways. It also appears to stimulate the production of adreno-cortical hormones and expend bronchial smooth muscles. In recent studies it has been shown to improve the efficacy of glucocorticoid treatments.

While the effectiveness of Cordyceps sinensis in the treatment of arrhythmias is relatively known, it mechanism is not. It has been suggested that by decreasing the permeability of the cell membranes to sodium and potassium ions, reducing the repolarization rate and the amount of free radicals available in the cell, that the fungus may treat arrhythmias.

Cordyceps sinensis has been shown the reduce hospital stay, clinical manifestations and the need for a respirator in patients suffering from chronic pulmonary heart disease. It has also been shown to improve respiration by decreasing the number of false neurotransmitters developed from an increase in phenylalanine and tyrosine.

The quality of life of patients suffering from chronic heart failure appears to be greatly improved by Cordyceps sinensis. Symptoms such as shortness of breath, fatigue, depression, anxiety, low stamina and poor self control were definitely better after the use of the fungus. It has been suggested that the mushroom can expand coronary vessels, increase blood flow to cardiac muscle and increase hypoxia tolerance. Further studies should be done to confirm these statements.

When Cordyceps is used in the treatment of atherosclerosis, it increases the immune response of red blood cells by increasing their activity in removing circulating immune complexes from the walls of the arteries. It also reduces the amount of cholesterol present in the blood and increases the phagocytic activity of white blood cells. It is worth mentioning again that increased cholesterol levels with increasing dosage of Cordyceps sinensis produced decreased aortic lipid deposits. This finding appears to indicate that the most important factor in the development of atherosclerosis is not the amount of cholesterol present in the blood but the inflammation process that develops in the arteries. As such one may wonder if cholesterol reducing drugs are the best course of action to treat atherosclerosis especially when one is aware of the side effects associated with them.

The impact of Cordyceps sinensis on chronic fatigue has in my opinion been rather exaggerated. It is mostly based on speculation and the observation that the fungus improves the immune system and that chronic fatigue syndrome may be linked to an unregulated immune system. The fact that healthy individuals feel less tired when working out when taking the fungus has also prompted researcher to conclude that the mushroom would indeed help with chronic fatigue syndrome as fatigue and low energy are the hallmark of this disease. Even if these observations appear promising they are not proof that Cordyceps does indeed help with chronic fatigue syndrome. Further study is definitely required.

Cordyceps has been known as a sexual tonic to the Chinese for almost two thousand years. The positive impact on libido, impotence and infertility (particularly male) is well documented. However the mechanism of action is generally not well understood.

Very few side effects are discussed in the research done on Cordyceps. An overwhelming majority of them actually do not mention any. However it is my experience that diarrhea is a rather common one and that larger doses may cause digestive upsets. Nausea and intense borborygmus may be experienced as well. An interesting side effect worth mentioning is the weight loss often accompanying patients taking the fungus, particularly in overweight diabetic patients.

As an aside, most research on Cordyceps sinensis available in the West is dismal at best. Almost all the articles used in this paper were found in China and translated in order to complete this project. It is sad to see that so little is done to bring this vast amount of information to the western practitioners and patients.

Overall Cordyceps sinensis' actions on the human body have been demonstrated even if the mechanisms associated with them are not always understood. However It seems that the effects of this amazing fungus are still not fully discovered and that the medical applications related to it are not all established. For this reason further study is essential. And as the west needs to understand how the mushroom works before it will fully integrate it into its medical treatment protocols, the need to study the biochemical mechanisms associated with each disease is of the utmost importance and should be researched diligently.

A final word: Cordyceps sinensis is one herb among many used by the oriental medical practitioner. It is an amazing herb that works even better when combined with other herbs. Do not attempt to treat yourself with it without first consulting a health care professional trained in Chinese herbology. As competent as you MD may be, his understanding of Chinese herbs is limited at best. For optimum health one needs both types of medicine. In my own practice, I always invite other medical practitioner to contact me in order to share what I do and thus provide the highest possible care for my patients. Expect no less from those who treat you.

# References

Books:

-       Bensky, Dan; Clavey, Steven; Stoger, Erich. Chinese Herbal Medicine Materia Medica. 3rd Ed. Eastland Press. Dong Chong Xia Cao. 2004. 770-773.

-       Lloyd, C.G. "Cordyceps sinensis, from N. Gist Gee, China". Mycological Notes. June 1918. 54. 766-767.

-       Pereira, Jonathan."Summer-Plant-Winter-Worm". New York Journal of Medicine and Collateral Sciences.1883. 1. 128-130.

-       Porth Mattson, Carol. Pathophysiology Concepts of Altered Health States. 5th Ed. Lippincott. 1998.

-       Thibodeau, Gary. A; Patton, Kevin. T; Anatomy and Physiology. 4th Ed. Mosby Inc. 1999.

-       Ross, Jeremy; Zang Fu. The Organ Systems of Traditional Chinese Medicine, Functions, Interrelationships and Patterns of Disharmony in theory and practice. 2nd Ed. Churchill Livingstone. 1985. 92-93.

-       Maciocia, Giovanni; The Foundations of Chinese Medicine. A Comprehensive Text for Acupuncturists and Herbalists. Churchill Livingstone. 1998. 77-87, 95-104.

-       Li, Peiwen; Cheng, Zhiquiang; Du, Xiuping. Management of Cancer with Chinese Medicine. Donica Publishing Ltd. Etiology and pathology of tumors. 19-27.

-       Halstead, Bruce W, Holcomb-Halstead, Terri L; The Scientific Basis of Chinese Integrative Cancer therapy including A Color Atlas of Chinese Anticancer Plants. North Atlantic Books. 2002. 71-78, 85-104, 198.

-       Berkow, Robert MD.1992. The Merk Manual Volume II Specialties. 16th Ed. Merk Research Laboratories. Human Immunodeficiency Virus (HIV) Infection. 141-148.

- Flaws, Bob; Sionneau, Philippe. The treatment of Modern Western Medical Diseases with Chinese Medicine, A Textbook and Clinical Manual.2nd Ed. Blue Poppy Press. 2005.

- Chen, Da-Can; Xuan, Guo-Wei; The Clinical Practice of Chinese Medicine: Lupus Erythematosus. People's Medical Publishing House. 2007. 13-35.

- Tierney Jr, Lawrence, McPhee, Stephen and Papadakis, Maxine. 2004 Current Medical Diagnosis & Treatment. 43st Ed. Lange Medical Books/Mc Graw Hill. 2004.

- Halpern, Georges. M. PhD. MD. Healing Mushrooms. Square One Publishers. 2007. 65-86.

- Li, Wei. L.Ac; Friedman, David. L.Ac. Clinical Nephrology in Chinese Medicine. Blue Poppy Press. 2003. 227-238.

- Zhang, Qingcai. Healing Hepatitis C with Modern Chinese Medicine. Sino-Med Institute. 2000. 25, 27, 36, 38-40, 79-81, 90, 94, 113.

- Becker, Simon; Flaws, Bob; Casanas, Robert. MD. The Treatment of Cardiovascular Diseases with Chinese Medicine. Blue Poppy Press. 2005. 3-19, 65-100, 137-214.

- Lyttleton, Jane. Treatment of Infertility with Chinese Medicine. Churchill Livingstone.2004.

- Lahans, Tai. Integrating Conventional and Chinese Medicine in Cancer Care A Clinical Guide. Churchill Livingstone. 2007.

- Halstead, Bruce W MD; Holcomb-Halstead, Terri L. The Scientific Basis of Chinese Integrative Cancer Therapy including A Color Atlas of Chinese Anticancer Plants. North Atlantic Books. 2002.

- Halpern, Georges M. PhD. MD. Cordyceps China's Healing Mushroom. Avery Publishing Group. 1999.

Articles:

- Hsu, Tai-Hao; Shiao, Li-Hua; Hsieh Chienyan; Chang, Der-Ming."A Comparison of the chemical composition and bioactive ingredients of the Chinese medicinal mushroom DongChongXiaCao, its counterfeit and mimic, and fermented mycelium of *Cordyceps sinensis*". Food chemistry 78.2002. 463-469.

- Dong, C. H; Yao, Y. J. "Nutritional requirements of mycelia growth of Cordyceps sinensis in submerged culture". Journal of Applied Microbiology. 2005. 99. 483-492.

- Li, S.P.; Zhang, G.H.; Zeng, Q.; Huang, Z.G.; Wang, Y.T.; Dong, T.T.X.; Tsim, K.W.K. "Hypoglycemic activity of polysaccharide, with antioxidation, isolated from cultured Cordyceps mycelia". Phytomedicine: International Journal of Phytotherapy & Phytopharmacology. June 2006.428-433.

- Lo, H.C.; Tu, S.T.; Lin, K.C.; Lin, S. C. "The anti-hyperglycemic activity of the fruiting body of Cordyceps in diabetic rats induced by nicotinamide and streptozotocin."Life Sciences. April 23 2004. 2897-908.

- Lo, H.C.; Hsu, T.H.; Tu, S.T.; Lin, K.C. "Anti-hyperglycemic Activity of Natural and Fermented Cordyceps sinensis in Rats with Diabetes induced by Nicotinamide and Steptozotocin". American Journal of Chinese Medicine. 2006; 34 (5). 819-832.

- Jin, Guohua; Zhao, Ming. "Compound Cordyceps sinensis Capsule to treat 31 cases with Diabetes Combined with Hyperlipidemia". Shaanxi Journal of Traditional Chinese Medicine. 2000. Issue 12.

- Liu, Feng; Zheng, Xiao. "Study of Cordyceps sinensis on Anti-Laryngeal Carcinoma". Journal of Norman Bethune University of Medical Science. 1993. 19 (1).

- Nakamura, Kazuki; Yamaguchi, Yu; Kagota, Samoti; Kwon, Young. Mi; Shinozuka, Kazumasa; Kunitomo, Masaru. "Inhibitory Effect of Cordyceps sinensis on Spontaneous Liver Metastasis of Lewis Lung Carcinoma and B16 Melanoma Cells in Syngeneic Mice". Jpn. J. Pharmacol. 1999. 79. 335-341.

- Nakamura, Kazuki; Yamaguchi, Yu; Kagota, Samoti; Shinozuka, Kazumasa; Kunitomo, Masaru. "Activation of In Vivo Kupffer Cell Function by Oral Administration of Cordyceps sinensis in Rats". Jpn. J. Pharmacol. 1999. 79. 505-508.

- Chen, Yu-Jen; Shiao, Ming-Shi; Lee, Shiuh-Sheng; Wang, Sheng-Yuan. "Effect of Cordyceps sinensis on the Proliferation and Differentiation of Human Leukemic U937 Cells". Life Sciences. 1997. 60 (25). 2349-2359.

- Wang, Jian; Zou, Wen. "A General Introduction of HIV/AIDS Treatment with Traditional Chinese Medicine in China". Virologica Sineca. 2007. 22 (6). 471-475.

- Lu, Weibo; Wen, Ruixiang; Guan, Chongfen; Wang, Yizhe; Shao, J.; Mshiu, M.; Mbena, E. "A Report On 8 Seronegative Converted HIV/AIDS Patients with Traditional Chinese Medicine". Chinese Medical Journal. 1995. 108 (8). 634-637.

- Gai, Ling; Song, Chunqing; Hu, Zhibi; Gai, Yun. "196 Summary of Research on Botanical Anti-AIDS". Foreign Medicine. Chinese Traditional Medicine Fascicle. 2001. 23 (3).

- Wu, Jian. Yong; Zhang, Qiao. Xia; Leung, Po. Hong. "Inhibitory effects of ethyl acetate extract of Cordyceps sinensis mycelium on various cancer cells in culture and B16 melanoma in C57BL/6 mice". Phytomedicine. 2007. 14. 43-49.

- Kinjo, Noriko; Zang, Mu. "Morphological and phylogenetic studies on Cordyceps sinensis distributed in southwestern China". Mycoscience. December 2001. 42 (6). 567-574.

- Montefiori, David C; Sobol, Robert W. Jr; Li, Shu Wu, Reichenbach, Nancy L; Suhadolnik, Robert J; Charubala, Ramamurthy; Pfleiderer, Wolfgang; Modliszewski, Ann; Robinson, W. Edward. Jr; Mitchell, William M. "Phosphorothioate and cordycepin analogues of 2', 5'-oligoadenylate: Inhibition of human immunodeficiency virus type 1 reverse transcriptase and infection in vitro". Proceedings of the National Academy of Sciences of the United States of American. September 1989. 86. 7191-7194.

- Yarnell, Eric, Abascal, Kathy. "Lupus Erythematosus and Herbal Medicine". Alternative and Complementary Therapies. February 2008. 9-12.

- Chen, Jong-Rern; Yen, Jeng-Hsien; Lin, Chun-Ching; Tsai, Wen- Junn; Liu, Wen-Jan; Tsai, Jih-Jin; Lin, Sheng-Fung; Liu, Hong-Wen. "The Effects of Chinese Herbs on Improving Survival and Inhibiting Anti-ds DNA Antibody production in Lupus Mice". American Journal of Chinese Medicine. 1993. 21 (3-4). 257-262.

- Chen, Yiping; Liu, Weizu; Zhao, Peizhu; Shen, Lingmei. "Initial Observation of 28 Cases with Chronic Kidney Failure Treated Mainly with Cordyceps sinensis". Shanghai Traditional Chinese Medicine Journal. 1984. 2.

- Xu, Cuiping; Chen, Kang. "Forty Cases of Chronic Renal Failure Treated by Combination of Chinese and Western Medicine". Henan Traditional Chinese Medicine. March 2008. 28 (3).

- Xu, F; Huang, J.B; Jiang, L; Xu, J; Mi, J. "Amelioration of cyclosporin nephrotoxicity by Cordyceps sinensis in kidney-transplanted recipients". Nephrology Dialysis Transplantation: letters. 1995. 10 (1).142-143.

- Chen, Yan. "Protecting Effects of Cordyceps sinensis for Renal Lesion Caused by Cisplatin with 42 Clinical case reports". Jiangsu Traditional Chinese Medicine. 2003. 24 (12).

- Guo, Zhaoan. "Research Status of Cordyceps sinensis in Treating Renal Failure in China". Chinese Journal of Integrated Traditional and Western Medicine in Intensive and Critical Care. August 1996. 3 (8).

- Wang, Ting; Wang, Yu-gang. "To explore the Effects of Cordyceps sinensis on Alcoholic Fatty Liver in Rats". Guide of Chine Medicine. October 2008. 6 (19).

- Yang, Zhao-xia; Shen, Wei; Dai, Dong-ling. "Effect of Cordyceps sinensis on Experimental Rat with Fatty Liver and Its Possible Molecular Mechanism". Chongqing Medical Journal. Sept. 2006. 35 (18).

- Liu, Yu-Kan; Shen, Wei. "Inhibitive effect of Cordyceps sinensis on experimental hepatic fibrosis and its possible mechanism". World Journal Gastroenterology. 2003. 9 (3). 529-533.

- Xu, Lieming; Liu, Chong; Liu, Ping; Zhu, Jianliang; Lv, Gang; Xue, Huiming; Shen, Songfa; Hu, Yiyang; Hong, Jiahe. "Treating posthepatic Cirrhosis with Extractum Semen Persicae plus Cordyceps sinensis Berk Sace: A Research on Pathology and Immunohistochemistry". Chinese Journal of Integrated Traditional and Western Medicine on Liver Diseases. 1994. 4 (1).

- Wang, Yaojun; Quan Qizhen; Sun Ziqin et al. "Therapeutic Effects of Cordyceps sinensis on Decompensated Cirrhosis". Hebei Medical Science. 1996. 2 (2).

- Liang, Huijing; Niu, Guoming; Guan Yangfang. "Cordyceps sinensis combined with Anti-HBs Positive Placenta to treat Hepatitis B (28 cases)". Shandong Medicine. 2006. 46 (31).

- Gong, Huan-yu; Wang, Ke-qing, Tang, Shi-gang. "Effect of Cordyceps sinensis on T lymphocyte subsets and Hepatic Fibrosis in Patients with Chronic Hepatitis B". Bulletin of Hunan Medical University. 2000. 25 (3).

- Zhou, Liangmei; Yang,Weizhu; Xu, Yuemin; Zhu, Qinyi, Ma Ziliang; Zhu, Tingrui; Ge, Xioayan. "Observation on Short-Term Curative Effect of Cultured Cordyceps sinensis for Chronic Hepatitis B". China Journal of Chinese Materia Medica. Jan 1990. 15 (1).

- Gai, Guozhong; Jin Shunji; Wang, Bo; Li Yongqing, Li Chongxian. "The Efficacy of Cordyceps militaris Capsules in Treatment of Chronic Bronchitis in Comparison with Jinshubao Capsules". Chinese New Drugs Journal. 2004. 2.

- Guan, Cai-hong; Liu Jin. "Effect of Cordyceps sinensis on Airway Inflammation and pulmonary functions of Rats with Chronic Obstructive Pulmonary Disease". Zhejiang Medical Journal. 2007. 29 (2).

- Qian, Haoyu. "Therapeutic Effects of Cordyceps sinensis on Chronic Obstructive Pulmonary Disease". Journal of Medical Forum. June 2004. 25 (11).

- Holliday, John; Cleaver Matt. "Medicinal Value of the Catarpillar Fungi Species of the Genus Cordyceps (Fr.) Link (Ascomycetes). A Review". International Journal of Medicinal Mushrooms. 2008. 10 (3).219-234.

- Wang, Ningqun; Jiang, Liangduo; Zhang, Xiaomei; Li, Zongxin. "Effect of Cordyceps sinensis Capsule on Airway Inflammation of Asthmatic Patients". China Journal of Chinese Materia Medica. 2007. 15.

- Du, Juan; Zhou, Jian; Chen, Bingrong; Hu, Gubing. " Influences of Cordyceps sinensis on Cellular Immune Function of Asthmatic Children". Zhejiang Journal of Integrated Traditional Chinese and Western Medicine. 2006. 16 (1).

- Fang, Yunchun; Han, Shuren; Xu, Lihua; Xi, Zhaoqing, Wang, Junmei; Feng, Qunxian. "Application of Compound Cordyceps sinensis Granules in Treating 65 Cases with Chronic Bronchitis". Journal of Nanjing College of Traditional Chinese Medicine. 1988. 1.

- Gong, Xiaojian, Ji, Hui; Cao, Qi ; Li, Shaoping, Li, Ping. "Antagonistic Effects of Extracts from Cultured Mycelium of Cordyceps sinensis on Arrhythmia". Journal of China Pharmaceutical University. 2001. 32 (3). 221-223.

- Yu, Zhou; He, Jianxin. "Therapeutic effect of Water Extract of Cordyceps sinensis on Triggered Arrhythmias". China Journal of Traditional Chine Medicine and Pharmacy. 1999. 14 (1).

- Wang, Jiangxian. "The Effect of Cordyceps sinensis on the Cure for Arrhythmias". Chinese Journal of New Drugs and Clinical Remedies. 1992 (5).

- Yang, Zhibing, Zhan, Chun; Xiao Lang ; et al. "Therapeutic Observation on 30 Cor Pulmonale Cases with Heart Failure Treated with Cordyceps sinensis". Chinese Journal of Integrated Traditional and Western Medicine. 2002. 12 (5).

- Xiao, Lang; Liang, Jing. "Therapeutic Observation of Cordyceps sinensis (Dongchongxiaocao) and Corbrin Capsule in the Treatment of Chronic Pulmonary Heart Diseases with Respiratory Failure. Tianjin Pharmacy. 2006. 6.

- Gu, Xuekui; Wu, Wei; Ju, Shaobin. "Application of Ginseng Cordyceps sinensis Soup in Treating 32 Cases with Chronic Congestive Heart Failure". New Journal of Traditional Chinese Medicine. 1995. 8.

- Hammer, Leon MD; Heffron, Robert MD; Leavy, Kathleen RN. "Inflammation in Atherosclerosis". Medical Acupuncture. 2003. 15 (2).

- Ma, Xiao; Guan, Yi-jun; Zhang, Wei. "The therapeutic Effect and its Mechanism of Cordyceps sinensis Compound on Experimental Atherosclerosis". Journal of Shangdong University (Health Sciences). Jan 2005. 43 (1).

- Yamaguchi, Yu; Katoga, Satomi; Nakamura, Kazuki; Shinozuka, Kasumasa; Kunimoto, Masaru. "Inhibitory Effects of Water Extract from Fruiting Bodies of Cultured Cordyceps sinensis on Raised Serum Lipid Peroxide Levels and Aortic Cholesterol Deposition in Atherosclerotic Mice". Phytotherapy Research. 2000. 14. 650-652.

- Xiao, Yi; Huang, Xi-zhen; Zhu, Jia-shi. "Randomized Double-blind Placebo-controlled Clinical Trial and assessment of Fermentation Product of Cordyceps sinensis (cs-4) in Enhancing Aerobic Capacity and Respiratory Function of Healthy Elderly Volunteers". Chinese Journal of Integrative Medicine. 2004. 10 (3). 187-192.

- Zhu, Jia-Shi MD. PhD; Halpern, Georges M. MD. PhD; Jones, Kenneth. "The Scientific Rediscovery of an Ancient Chinese Herbal Medicine: Cordyceps sinensis Part I". The Journal of Alternative and Complementary Medicine. 1998. 4 (3). 289-303.

- Yuan, Hongfen; Zhu, Jianping. "Observation of Clinical Effects of Cordyceps sinensis in Treating Seminal Emission". Contemporary Medicine. June 2008. 142.

- Huang, Bu-Miin; Hsu, Chih-Chao, Tsai, Shaw-Jeng; Sheu Chia-Chin, Leu, Sew-Fen. "Effects of Cordyceps sinensis on testosterone production in normal mouse Leydig cells". Life Sciences. 2001. 69. 2593-2602.

## Online sources:

- http://pharmaceuticalmushrooms.nwbotanicals.org/lexicon/cordyceps. 12/19/2008

- http://www.umm.edu/ency/article/000821.htm.%20. 12/12/2008

- Lu, L.; "Study on effect of Cordyceps sinensis and artemisinin in preventing recurrence of lupus nephritis.(Author Abstract)." Alternative Medicine Review. Thorne Research Inc. 2003. *HighBeam Research.* 30 Dec. 2008 <http://www.highbeam.com>.

- http://en.wikipedia.org/wiki/liver.%20. 12/19/2008

- http://www.engenderhealth.org/res/onc/sexuality/dysfunction/pg2.html.%20. 03/15/2009.

- Kaptchuk, Ted. "On Nature, Science and their Dangers". www.tcmnordic.sc/HerbalCrossroadsSeptember2005.doc. 03/15/2009.